"Cool is as important as it is elusive. People want to find it and brands want to be it, but what is it and why do we all care so much? *Cool* probes the far reaches of our brain to answer these questions, shining a light on the essence of cool and the fundamental motivations that make us all human."

—Jonah Berger, Associate Professor of Marketing, Wharton School, University of Pennsylvania, and bestselling author of *Contagious*

"Both a sweeping history and a scientific exploration, *Cool* charts the evolution of an ineffable concept that, whether or not we realize it, influences our decision making every day. Reading this book can't make you cool, but it can give you the tools to figure out why cool matters."

—Richard Florida, Director of the Martin Prosperity Institute, University of Toronto, and bestselling author of *The Rise of the Creative Class*

"Steven Quartz and Anette Asp expose a mystery that plagues us all—spending—and do so by uncovering the biological roots that guide our desire for status while following ancient rules that kept our evolutionary forebears alive. They remind us that forces that drive our modern habits were put in place long before there was anything modern. *Cool* is a delightful book that will inspire discussion."

—Read Montague, Director of the Human Neuroimaging Laboratory, Virginia Tech Carilion Research Institute, and author of *Your Brain Is (Almost) Perfect*

"Quartz and Asp certainly believe that [the quest for cool] can lead us to meaningful social experiences . . . What's cool? *You* decide—and then watch it enhance your well-being and relationships."
—Kevin Evers, *Harvard Business Review*

"Intriguing . . . What's really fascinating about this book is its argument as to how and why 'cool,' as a status-seeking strategy, changed from being a prerogative of the rich and well-born to being open to anyone—and in so doing spawned the much more diverse, fragmented and pluralistic culture we presently enjoy in the West."
—Robert Collison, *Toronto Star*

STEVEN QUARTZ AND **ANETTE ASP**

COOL

Steven Quartz is a professor of philosophy and cognitive science and the director of the Social Cognitive Neuroscience Lab at the California Institute of Technology. He is the coauthor of *Liars, Lovers, and Heroes* and lives in Malibu, California.

Anette Asp is a political scientist, public relations and communications professional, and pioneer in the field of neuromarketing. She is a former project manager at the Social Cognitive Neuroscience Lab at the California Institute of Technology and is currently the communications manager of a leading telecommunications company. She lives in Stockholm, Sweden.

COOL

COOL

HOW THE BRAIN'S HIDDEN QUEST FOR COOL DRIVES OUR ECONOMY AND SHAPES OUR WORLD

STEVEN QUARTZ and ANETTE ASP

FARRAR, STRAUS AND GIROUX   NEW YORK

Farrar, Straus and Giroux
120 Broadway, New York 10271

Published in 2015 by Farrar, Straus and Giroux
First paperback edition, 2016

The Library of Congress has cataloged the hardcover edition as follows:
Quartz, Steven.
 Cool : how the brain's hidden quest for cool drives our economy and shapes our world /
Steven Quartz, Anette Asp.
  pages   cm
 Includes index.
 ISBN 978-0-374-12918-7 (hardback) — ISBN 978-1-4299-4418-2 (e-book)
 1. Consumer behavior.   2. Consumption (Economics)—Social aspects.   3. Brain.
I. Asp, Anette, 1976–   II. Title.

HF5415.32 .Q37 2015
306.3—dc23   2014031273

Paperback ISBN: 978-0-374-53593-3

Designed by Jonathan D. Lippincott

Our books may be purchased in bulk for promotional, educational, or business
use. Please contact your local bookseller or the Macmillan Corporate and
Premium Sales Department at 1-800-221-7945, extension 5442,
or by e-mail at MacmillanSpecialMarkets@macmillan.com.

www.fsgbooks.com
www.twitter.com/fsgbooks • www.facebook.com/fsgbooks

CONTENTS

COOL

THE CONSUMPTION MYSTERY

Reflections of palm trees sway across Gucci's temple-like storefront in the brilliant afternoon sun. The glistening industrial storefront next door bears neither name nor address, evoking Prada's minimalist cool. Inside, a parade of mannequins is arranged with military precision, their averted gaze heightening their aloofness to passersby. One store down, the sun warms $15,000 Fendi bags, the scent of sumptuous leather in the air, and at Bijan, $20,000 silk suits wait patiently for those who have an appointment. Eight-hundred-dollar jeans, carefully torn at the knees and splattered with paint, grace the window display at Dolce & Gabbana. Above them hangs a sign crafted by some marketing consultant reassuring you that $800 for a pair of jeans is money well spent, as these jeans will make you even cooler than you already are. Rodeo Drive in Beverly Hills may seem an unlikely place to find Caltech scientists doing fieldwork. Sometimes, though, clues to the deepest mysteries about ourselves come from unlikely sources.

There's something odd about the fact that a row of stores is among the most famous tourist attractions in the world. On this typical summer afternoon, most of the people sauntering up and down Rodeo Drive are posing in front of the storefronts for souvenir pictures, panning the street with their video apps, and pressing their noses up to the display windows. Since they aren't actually

shopping, the street's attraction must have some largely unacknowl-
edged ritualistic flavor to it. To an anthropologist from another planet,
we suspect, the throngs of tourists would be as exotic and mysterious as any congregation of premodern humans chanting and dancing
around a campfire some warm, starry night on a far-off savanna.

What brings these people to Rodeo Drive? What is its allure?
Their mood offers a clue. Watching them strolling, gawking, and posing: You can't help but notice that they are almost giddy, their heads
no doubt filled with fantasies stemming from the modern-day fairy
tale *Pretty Woman*, and the magical transformative power of this
place. For adults, this—and not some amusement park an hour
south—looks like the happiest place on earth. It is, of course, more
than amusement. It is aspirational. We are so intimately familiar
with the link between happiness and consumption that it may never
occur to us that Rodeo Drive is something like a shopper's Canter-
bury or a consumerist Mecca, if you'll indulge us in some mixed
metaphors. That is, Rodeo Drive's lure lies in something abstract, in
its distillation of the very essence of consumerism: the promise that
personal happiness can be found by consuming more than one needs.
To that alien anthropologist, the people on Rodeo Drive must seem
like pilgrims who have traveled countless miles to let the opulence
of its offerings and all that goes with the promise of consumerism
wash over them like a balm.

We are all consumers.[1] And we all, more or less, live by consumerism's creed that our consuming is linked to our happiness (in a
recent poll, only 6 percent of Americans said that money can't buy
happiness).[2] When someone says money can't buy happiness, they
typically mean buying "stuff" can't buy happiness. But consumerism is more than just buying stuff. It also makes possible a dizzying
array of experiences and lifestyles. Elizabeth Gilbert's bestseller
Eat, Pray, Love may have gained Oprah's attention as a woman's
search for meaning, but Gilbert's yearlong travels—from savoring
the cuisine of Italy to taking yoga lessons in India—was an ode of
sorts to consumerism and a lifestyle it made possible. In fact, "things"
and "experiences" are often so interwoven that we can't really separate them. Two tickets to your favorite baseball team's game are
things, but taking your child to the game might be an unforgettable

experience. A bicycle is a thing, but it might offer the experiences of an annual bike trip through the wine country with friends. It might even offer a weekly ride with a local club, travel to take part in races, and a whole cycling way of life. *Being a cyclist*—a lifestyle made possible by consumerism—might soon start to define who you are.

Just think of how your own pattern of consumption conveys who you are to yourself—and to others. For according to consumerism, without your clothes you are more than naked. You are meaning-less. This is because in a consumer culture things live a double life, both as material objects and as symbols or signals with meanings, both explicit and unrecognized, that communicate values, identi-ties, aspirations, and even fears. All these add up to our lifestyles, made possible by consumerism. Indeed, according to some social critics, it is through the world of commodities that our social world reproduces the social categories that structure our personal identi-ties and give form to the social order.

If you want to put the deeply symbolic nature of material things to the test, just drive a Hummer to an environmental meeting or a Prius to a NASCAR race and wait for the reaction you get. The green urban-hipster values of the Prius don't play so well to the NASCAR crowd, while the Hummer's embodiment of a middle fin-ger to the environment has made it a target of "ecoterrorists." We are awash in these signals, from the cars we drive and the clothes we wear to the brand of hand soap beside our kitchen sinks (and the kitchen sink itself, for that matter). Many of these signals were shaped by our evolutionary past and speak to our brains as ancient symbols, below the level of our awareness. They motivate and guide our behaviors in ways we rarely acknowledge and sometimes even vehemently deny.

There is little or no avoiding this world of goods, symbols, and signals. Even the self-proclaimed "anti-consumers" among us typi-cally end up being just alternative consumers. Consider, for exam-ple, movements like the Simple Living Network, an anti-consumption group (now defunct) that offered to provide resources for learning to do more with less. Without a hint of irony, its website peddled Simple Living bumper stickers, T-shirts, banners, books, posters, flags, buttons, magnets, note cards, and a veritable laundry list of other goods. And the anti-consumerist organization Adbusters is busy supplying its supporters with its own $125 in-house brand of sneakers, which are no longer clothing but—so the marketing proclaims—have been transformed into rebellious anti-corporatist "tools for activists." And you thought a shoe was just a shoe. Even consider the avowedly nonconsumer, off-the-grid Amish. Recent times have seen more and more Amish trading in the horse and plow for high-paying factory jobs and enjoying the fruits of their labor by dining out regularly and even vacationing in Florida. So popular have winter Amish treks to Florida become that an entire Amish vacation community, Pinecraft, has sprung up just outside Sarasota, where vacationing Amish enjoy deep-sea fishing, parasailing, and shuffleboard. An Amish paradise indeed.

Once synonymous with the West, consumerism has spread widely across the world and today coexists with political and religious climates once strictly antithetical to it. A fitting sign of consumerism's increasingly global reach is the fact that the largest shopping mall in the world is the New South China Mall, more than twice the size of the Mall of America, the largest mall in the United States. In fact, all of the world's ten biggest malls are in Asia or the Middle East. What may appear to be worlds away is oddly connected to us by the language of consumption. While you may at first struggle to find common experiences with someone living in Chengdu in Southwest China or in Sharjah in the United Arab Emirates, the globalization of consumerism provides common touchstones. You would likely be able to share with them the experience of stopping at Starbucks in the mid-afternoon to enjoy a mocha Frappuccino.

Given how central consumerism is to our lives, and given its growing reach, you might suppose we'd all have a good grasp on our reasons for consuming. But when we started asking people in 2003 why they consume, we found that they had a hard time answering. We were using new brain-imaging technologies to peer into their brains as they made consumer decisions, and we supposed their impressions might help us interpret the brain scan results. Soon, however, we discovered that the brain scans were illuminating "the why of buy" in ways that went far beyond the subjects' introspection.[3] Brain imaging was opening a window into the unconscious brain, which, we were discovering, drives most of our consumer behavior.

Then something unexpected happened. In the spring of 2004, we were conducting a brain-imaging experiment involving "cool" and "uncool" products. It started out as a somewhat lighthearted look into what we thought was an interesting but not central part of our economic life. We didn't expect "cool" to be a game changer. But as we delved into deciphering the brain-imaging results, we realized that they didn't fit with the popular theories of consumer behavior that economists, psychologists, and sociologists had proposed. Looking inside the brain could finally answer why we consume, but it would require a new understanding that would take us a decade of effort to work out, and that would force us to rethink many of our most basic assumptions.

This new brain-based understanding exposes many of our deepest beliefs about consumerism as myths. Human consumption, it turns out, stems from the very same sources as our moral behavior. Our brain studies have also revealed how a special kind of consumption helped to solve an incredibly basic social problem, which we refer to as the Status Dilemma. That solution was the "rebel cool" that emerged in the 1950s, a new, oppositional style of consumption. Another kind of cool consumption, which we call "Dot-Cool," emerged in the 1990s. As much as we may be biased to disdain consumerism, the emergence of these new kinds of consumption forces us to seriously reconsider consumer behavior in a new light.

In *Cool*, we present this new understanding of why we consume, how cool consumption emerged as a prime driver of the global economy, and how cool consuming shapes our world. Our view draws on the emerging science of "neuroeconomics" and a view Steve first articulated in the 1990s, "cultural biology."[4] Neuroeconomics is a field that is fast uncovering our brain's hidden economic life. Like much work in this field, our work challenges the traditional economic conception of the consumer, known as *Homo economicus*. This hypothetical character is a bit like Mr. Spock from *Star Trek*: consuming for him is like figuring out a math game involving rational calculation. He would never buy a new shirt just because a salesperson complimented him when he tried it on. In fact, he makes his economic decisions as though he were the only person in the world. In our experiments involving "cool" products, in contrast, we found that asking people to merely look at cool products sparked a pattern of brain activation similar to what we see when we ask people to do social tasks, such as imagining themselves in a social situation or interacting with others directly. This is a tantalizing clue that part of the economic value of these products lies in the brain's mostly implicit estimate of how they impact our social identity. But deciding if you like a cool product or not is supposed to be a question of economics! After all, you typically hear about "consumers" on the business news, not the celebrity news. If we like cool products because they somehow tap into our social brain, then consumption doesn't fit the traditional, "rational" economic model.[5]

Cultural biology also places a premium on understanding our social life as the interplay between instincts, rooted in ancient neural structures whose design can be traced back to insects, and our capacity for cultural learning. The human brain develops through a prolonged and rich interaction with the environment over the first two decades of life. Nowhere is this rich interaction more pronounced than in the region of the human brain that expanded most during human evolution, an expansion that made possible our extraordinary sociability and also happens to be activated by cool products, as mentioned previously. That is, the brain region that blends our economic decisions and our social identity both grows the most during development and takes the longest to develop.[6] This is not a coincidence. We learn how to associate products with our social identity and then how to use those products to signal what we're all about to other people. This latter ability emerges during adolescence as these brain regions develop—it's one reason why teenagers become so concerned about their social identity and communicate it to others through their burgeoning lifestyle choices.

Because of this complex biological and cultural interplay, cultural biology's view of consumption suggests that trying to understand human nature, including consumption, exclusively through the lens of evolutionary psychology (popularized in such books as Steven Pinker's *How the Mind Works*) is a mistake. Evolutionary psychology views human nature as the result of hardwired brain circuits that owe little of their function to culture, seeing us as essentially Stone Age creatures out of place in a modern world. On the contrary, cultural biology emphasizes the *interplay* between a flexible brain and culture, in which culture helps to *build* the brain and its functions. We suspect this is absolutely crucial if we are ever going to understand how a rapidly changing consumer culture plays a role in reshaping our beliefs and behavior, which have transformed profoundly over the last fifty years, as we will see in chapters 7 and 8.

As important as the social environment's role is in shaping our brain, it's not the whole story. Our brain has been primed for consumption throughout its evolutionary history. In fact, what makes modern consumption such a powerful force in our lives is that it builds on desires and motives that are etched very deep into our brains. In other words, it is part of our nature to consume. As we look into the ancient forces that shaped the modern brain and our consuming nature, we'll discover that like our closest genetic relative, the chimpanzee, we instinctively seek status. While some consumer critics point to our status impulses, they misdiagnose status as purely a competition for individual distinction that builds fences between people. Viewed in the appropriate evolutionary framework, our status instinct, we'll discover, is rooted in our brain's most basic affiliative impulses, which makes consumption more about building bridges between people. We also share with the chimpanzee a rebel instinct that makes us instinctively resent being subordinated. To our knowledge, no one has explored how this rebel instinct figures in our consumption. We thus examine how modern consumption didn't require the invention of new needs, but instead builds on these two instincts. This is a reason why consumption spreads like wildfire whenever conditions allow. Indeed, it took root quickly even among the hunter-gatherer Tsimané, who live in some of the most remote regions in the world, as soon as discretionary income became available and long before advertisements or other social influences appeared on the scene.[7]

A major barrier to understanding consumption is the idea that our status concerns are artificial, or worse yet, pathological. To our thinking, this is a historically monumental mistake, one that has resulted in decades of misleading consumerism critiques. Once we recognize the biological reality of consumer motives—the status instinct and the rebel instinct—and understand the critical role they play in our lives, the prescription to deny them becomes about as feasible—and right-minded—as the Victorian demand for chastity. Indeed, once we recognize that these instincts are a legitimate element of being human, we'll see cool consumption in a new light, as a solution to the Status Dilemma.

The view that status seeking is artificial, imposed by an unjust and crass society, remains so pervasive that it's worth examining in more detail. The most famous representative of this idea is *Homo sociologicus*. Whereas *Homo economicus* is an asocial creature, *Homo sociologicus* is created almost entirely by society. Modern human needs, the story goes, are essentially created by society as we take on the roles, and their attendant desires, that society makes available to us. It is a view most often traced to the eighteenth-century philosopher Jean-Jacques Rousseau and his notion of the Noble Savage. According to Rousseau, in a natural state we have few needs, but under the influence of society we develop artificial pride (*amour-propre*) that drives us to compare ourselves with others and to depend on their good opinion for our well-being (status). Civilization is thus a corrupting force. To Rousseau, material progress thwarts genuine human bonds of friendship and authenticity and replaces them with jealousy and artifice. The legacy of this view was the proliferation of critiques, ranging from Karl Marx's to that of the French postmodernist Jean Baudrillard, asserting that human needs are created by the system of production to drive the engine of capitalism.[8]

The basic idea here is that consumerism relies on instilling false needs in us to make us believe that our happiness depends on consuming. According to Alain de Botton, for example, status concerns and social hierarchies are constructed by a consumerist culture. By creating aspirations based on false needs, he believes, society creates a painful "status anxiety" within us.[9] This status anxiety is, he claims, entirely artificial. Yet once it has us in its grasp, it creates a desire to consume as a way to alleviate our pain. The result is that the apparent wealth of our society actually *impoverishes* us, as it creates unlimited expectations that leave us perennially unsatisfied. The lesson is that "Rousseau's naked savages had few possessions. But, unlike their successors in their Taj Mahals, they were at least able to feast on the great wealth that comes from aspiring to very little."[10]

In essence, then, *Homo sociologicus* consumes because he has been manipulated to do so by society. He is something of a passive dupe. He doesn't really choose or act in any meaningful way, but rather walks in lockstep with the demands of consumerism. This notion of modern consumption became the dominant explanation among sociologists for the rise of modern media, marketing, and advertising. The trouble began with modern production. As mass production and other forms of industrialization made the sheer numbers of products greater and greater, manufacturers faced a potential crisis of oversupply. To avert this crisis, the argument continues, they needed to create entire classes of consumers for their goods. To do so, they turned to a fledgling marketing and advertising industry, which made two key innovations. The first was the modern advertisement, developed with the help of the leading psychologists of the day, such as John Watson, one of the pioneers of American behaviorist psychology, who became an ad executive at J. Walter Thompson after an affair forced him from academia. The second innovation was new media to communicate those messages—mass media in the form of radio, movies, and ultimately television. As proponents of the "passive dupe" theory point out, the rise of modern advertising occurred during the largest economic expansion in U.S. history, the post–World War II boom. It was during this period that the notion that modern society produces a false consciousness reached its zenith, as it was tempting for many critics to place the blame for mass consumerism solely on mass media as instruments of manipulation.

The most influential exposé of new advertising techniques was Vance Packard's *The Hidden Persuaders*, which warned that mass media were using mind control techniques to embed false needs in listeners. Packard charged, among other things, that *The Howdy Doody Show* was subverting parental authority. Although this explanation of postwar America's consumerism remains popular, it's wrong. Postwar American consumerism was the result of one of the most ambitious periods of governmental economic planning in history, in which roles as consumer, citizen, and patriot intermingled.[11] Even so, the suspicion that consumerism is the product of media-driven manipulation remains widespread, as in Noam Chomsky's propaganda model of the media and their role in manufacturing consent.[12]

There's a variant of this view that is also important to highlight. We'll call him *Homo barbarus*. He is the creation of one of the most influential writers on consumption, the American economist Thorstein Veblen. In a hugely significant book, *The Theory of the Leisure Class*, Veblen likened our consumption to the barbarian's exploits of war. The driving force was neither rational utility maximization nor passive manipulation. It was irrational status competition. Although Veblen regarded his 1899 book as a work of economic analysis, his heavy use of satire, fanciful anthropological musings, and skewering of the rich resonated with the public, making his book a bestseller.

Contributing to the wide interest in his work was the fact that Veblen provided an intuitive explanation for the excesses of the Gilded Age in which he wrote. The previous few decades in the United States had witnessed some of the most rapid economic expansion in the history of the world. In the United States, which leapfrogged Britain as the greatest manufacturing nation in the world during this time, it was also the era of the tycoon, exemplified by John D. Rockefeller, J. P. Morgan, and Andrew Carnegie. Many of these tycoons, such as Carnegie, embodied both the American dream of "rags to riches" and the ruthlessness of the "new man of business." Veblen was fascinated by both these aspects of the American tycoon, as well as by the extraordinary lengths to which this new class of "businessmen" went to display their wealth. He melded these elements together into an anthropological account of the new businessman that reduced his toiling to the barbarian exploits of war. Veblen suggested that prehistoric human societies had passed from a stage of peaceable savagery to a barbarian warrior stage. Echoing sentiments akin to Rousseau's image of the Noble Savage, Veblen suggested that when humans became predatory, warlike emulation and "invidious comparison" emerged as prime motivators. Like Rousseau's *amour-propre*, Veblen's "invidious comparison" drives men to compete against one another for prestige, corrupting natural virtues and turning daily life into ceaseless contests of predatory exploits. (We say "men" here because Veblen believed these contests reduced the women of his day to trophies of these preda-

tory exploits—housewives were the modern equivalent of the captive women of war. Although Veblen did not coin the term "trophy
wife," the essential elements of the concept are there in his view.)
Veblen believed that the main difference between prehistoric bar-
barians and the businessmen of his day was the type of trophies
they displayed, making consumer goods little more than the modern signs of predatory gains. So if we were to apply Veblen's views to
our own era, just as a barbarian might raise the severed head of his
foe for all to see his might, today we flash a Rolex for all to see.

Fundamental to Veblen's analysis was his moral disapproval of
conspicuous consumption as inevitably wasteful, irrational status
seeking, which fueled his satirical attack on the leisure class. Wasteful consumption, to him, was any kind of consumption that didn't
serve a necessity of life, or did not "serve human life or human well-
being on the whole." As for Rousseau, irrational and wasteful consumption depends on drawing a distinction between genuine and
false human needs, and also on the implication that consumption is
antithetical to satisfying central human needs. From academics to
advice columnists, many of those discussing consumption today
likewise morally disapprove of it. For such critics, consumer culture
is deeply destructive. It seems unsustainable in a world of limited
resources. It erodes community. It hurts democracy by turning citizens into consumers. It makes us narcissistic. It causes a spiritually
deadening materialism.

The upshot is that many discussions of consumer culture today
are deeply tinged with moralism and focus more on condemning it
than on understanding it. The result is a void in our understanding
of a basic force shaping our world. Consider that the most basic political dynamic of the twentieth century was the competition between capitalism and communism over which system of production
could better satisfy the consumer needs of its citizens.[13] One of the
most iconic moments of the Cold War—the 1959 kitchen debate
between vice president Richard Nixon and Soviet premier Nikita
Khrushchev—took place in the model kitchen of a prototypical
American family home designed to showcase to Soviets the consumer goods available to the average American family. Khrushchev
flew into a profanity-laced tirade over Nixon's boast that American

technology had produced affordable dishwashers, washing machines, lawn mowers, cosmetics, hi-fi's, TV dinners, cake mixes, supermarkets, and convertibles. Khrushchev dismissed such things as trivial luxuries, but ultimately the inability to satisfy the consumer demands of its citizens was a force behind the Soviet Union's collapse. During the decade following that collapse, the Russian consumer market doubled in size and then doubled again in the five years after that. In China, party officials have pushed aside a puritan communism, with its official egalitarian and restricted consumption, in favor of a pragmatic consumerism that underlies one of the world's fastest-growing economies.[14]

With the fall of communism, the new clash of civilizations is between consumerist countries and consumerism's foes—Jihad vs. McWorld, as the political theorist Benjamin Barber dubbed it.[15] Why, then, does consumerism remain such a mystery? We suspect the reason lies in the fact that historically, the central problem for economies has been production rather than consumption.[16] Simply put, when your world is one of scarcity and famine, you don't need to make an argument for the desire to consume. Instead, the problem facing most people for most of history was how to expand production.[17] For that reason, the most historically influential political and economic theorists focused on the problems of production.

As solutions to those problems were worked out, through such ways as major advances in production technologies, a strange thing happened to consumption. Whereas theorists saw production as a good thing, and associated it with virtuous traits, they thought consumption was morally suspicious. Consider, for example, the German sociologist Max Weber's classical theory of the emergence of capitalism.[18] According to Weber, the rise of capitalism depended on two confluent factors: a work ethic stemming from the Puritan virtues of self-discipline, delayed gratification, and restraint, and a Protestant asceticism that shunned the accumulation of worldly goods. Together, Weber argued, these values created the work-and-save dynamic of modern capitalism. Consumption had long been a morally problematic notion: of the seven deadly sins, five are sins of consumption—pride, envy, gluttony, lust, and greed. Yet it was the stark moral contrast between production and consumption in Weber's theory of capitalism that provided a foundation for contemporary views of consumption as morally problematic.

This contrast is nowhere more striking than in the Harvard sociologist Daniel Bell's classic 1976 work, *The Cultural Contradictions of Capitalism*. According to Bell, capitalism underwent a profound transformation in the early twentieth century as it shifted from what Bell regarded as its rational focus on production to an irrational, anti-intellectual, promiscuous, and hedonistic emphasis on consumption. The driving force of this transformation, according to Bell, was the new bohemians, who flocked to places like Greenwich Village in New York City in search of unconventional lifestyles and sexual liberation. While the new bohemians' rejection of Puritan morality might at first appear threatening to a capitalism based on self-restraint, it was in fact assimilated into a "new capitalism" that was now based on hedonism. According to Bell, however, consumption-based capitalism is inherently unsustainable because hedonistic consumption rewards instant self-gratification, whereas production depends on hard work and delayed gratification. Such 1970s narratives of America's decline also latched onto narcissism as the sickness of their age.[19] The social critic Christopher Lasch charged that consumption was just narcissism run amok, a charge that is still echoed today.[20] As we'll see, it wasn't the first time cultural critics found a convenient bully pulpit in psychoanalysis to make their jeremiads appear scientific.

The widespread moral disapproval of consumerism got a boost from hugely influential findings dubbed the Easterlin paradox.[21] In 1974, the economist Richard Easterlin examined the question of whether economic growth increases happiness, and concluded that it didn't. Richer people were happier than poorer people in the same country. But overall, people in wealthier countries seemed no happier than those in poorer countries. All that seemed to matter was *relative* income. In fact, economic growth in a country didn't seem to result in any gains in happiness. Money, it seemed, didn't buy happiness. These results suggested that people's happiness only depends on how they fare compared with other people within their country—keeping up with the proverbial Joneses is all that matters. People in rich countries compare themselves to the Joneses in their country, and people in poor countries compare themselves to the Joneses in their country. The problem with economic growth is that increases in absolute wealth don't change your relative standing. Imagine, for example, if overnight we doubled the income of everyone in your country. You'd be right where you were yesterday in terms of keeping up with the Joneses. Even worse, the argument continues, an increase in absolute wealth might just be speeding up the "hedonic treadmill" to make you even less happy if you have to work longer hours to earn that increase in salary. This was a stunning rebuke of developmental economics and its central idea that increasing wealth should be the goal of economic policy. And it fit perfectly with the anti-consumption view that money can't buy happiness.

Our concern with relative income means that what we care about is status—our relative rank in society. Our acute concern with our relative place in society creates the Status Dilemma. That makes status a limited and fixed resource. Increases in everyone's absolute income don't add any more status, because status is all about relative income. The only way to gain status is for someone else to lose some (status is a zero-sum contest). The reason why Veblen's theory remains so influential today lies in the monumental move he made to link our status to our conspicuous consumption. As the Cornell University economist Robert Frank argues, consumption has become a status game, which means our status depends on how much we spend in a consumption arms race against our neighbors.[22] Just like two countries locked in a struggle neither one really wants to be in, according to this view we seem stuck in a senseless consumption race.

Virtually all critics of consumerism rely on some variant of the Easterlin paradox. It's the "incriminating evidence" supporting anti-consumption intuitions. In fact, when we began our work on consumer decision making, we assumed it was true and so viewed consumerism with a jaundiced eye. As compelling as the arguments appear—and as much as we may want to believe them—more recent analyses reveal there's no Easterlin paradox.[23] Using data from 140 countries, economists concluded that richer countries are significantly happier overall than poorer countries. As countries get richer, their citizens get happier. Absolute income matters after all.[24] Countries with the greatest economic growth have the highest levels of happiness. Indeed, global well-being has been rising since the 1960s at a rate that mirrors economic growth.

Then came a study by the psychologist Ed Diener, one of the pioneers in the science of happiness and its measurement.[25] Diener and his colleagues showed that the Easterlin paradox relied on a rough way of estimating income that was fraught with perils. Using more accurate measures, the paradox didn't just disappear: it was turned on its head. Rising incomes led to increased scores in life evaluations, as well as reports of more positive feelings and fewer negative feelings.

What's more, Diener's group discovered a critical link between rising income and happiness: material possessions. When more income translated into more purchasing power, people's happiness, financial satisfaction, and optimism increased. And this wasn't a fleeting effect, contradicting the claim of anti-consumerists that consuming results in at best only a brief uptick in happiness. The happiness from rising income was an enduring one.

The pervasive idea that we were happier before consumerism, echoed by such influential anti-consumer writers as Naomi Klein, turns out to be simply untrue.[26] So why is it such a compelling narrative? As the historian Arthur Herman chronicled in *The Idea of Decline in Western History*, the narrative itself isn't new. So-called declinism, especially prognostication of the inevitable decline of capitalism, has been a central theme in social thought for the last 150 years. Declinism—the belief that things were better in the past—has such a hold on us in part because our brains don't remember the past as it really was. We love to reminisce about the good old days, when movies and television were still good, the country was on the right track, and so on. In poll after poll, people think just about everything was better in the past.[27] But when researchers actually put it to the test, they discover that we remember things much more positively than we experienced them at the time.[28] It's called rosy retrospection and the nostalgic bias, and it's built into our brain.

Such biases have a strong hold on us. For example, when Steven Pinker presented mountains of data demonstrating that the rise of capitalism has led to less—not more—violence, contributing to the long historical decline of human violence, people howled with incredulity.[29] The very idea that things could be getting better just couldn't be right. So consider the following. In the 2014 Gates Foundation Annual Letter, Bill Gates opens by noting that "by almost any measure, the world is better than it has ever been."[30] Poor countries aren't doomed to stay poor. More than a billion people have risen out of extreme poverty. Foreign aid isn't a big waste of money, and saving lives doesn't lead to overpopulation. As the Swedish statistician Hans Rosling demonstrates in four minutes of stunning data animation, the last two hundred years have witnessed a remarkable worldwide shift as country after country has risen out of poor and unhealthy conditions to become healthy and wealthy.[31] The spread of women's reproductive rights has brought about a remarkable decline in the birthrate. The declinist narrative about consumerism impedes the very possibility of understanding it. We'll challenge the view that consumerism makes us unhappy and leads inevitably to the Status Dilemma. Indeed, we'll suggest it solves it.

The breakdown of the Easterlin paradox suggests that the needs driving consumption may not be so false after all. Consumption isn't contrary to human nature, as the influential "false needs" and "manipulation" views we traced above claim. Using the insights of neuroeconomics and cultural biology, we'll see that consumption stems from a status *instinct*. Demolishing the notion that status is a false, fabricated, or unnatural need, an understanding of our evolutionary past reveals why it's among the deepest-rooted human motives. A strong demand for status is built into our nature. But what exactly is status? How do we get status? And how much status is there, and what happens when there's "not enough" to go around? These are fundamental, but neglected, questions.[32]

The status instinct drives our emulation, jealousy, or envy of "higher-ups" and leads to *emulation consumption* (this bears some resemblance to the kind Veblen chronicled, though we'll discover major differences when we look at its evolution). But the status instinct and emulation consumption are only part of the story.

In the 1950s, consumption in the United States began to change radically. Cool emerged as an oppositional norm, a rejection of "higher-ups" and their traditional status system, to play a central role as an economic, social, and political force reshaping culture. In particular, cool soon drove a new kind of consumption—*oppositional consumption*—by invoking the rebel instinct. We refer to this new kind of consumption as rebel cool.

Although we often think of cool as a rebellion against consumption, rebel cool assimilated easily into consumerism to create new routes to status—new lifestyles whose values differed from traditional status. The fact that rebel cool came along to revolutionize consumerism during one of the greatest periods of economic growth and rising living standards in our history is extraordinarily intriguing. Growth in absolute income and a burgeoning consumerism led to an increase in the total amount of status. But we'll suggest that it required rebel cool as an oppositional force to bring down the traditional barriers to new lifestyles, barriers that included racial and gender discrimination and social institutions designed to maintain the status quo. As oppositional cool consumption emerged, its new lifestyles diversified and expanded the routes to status, washing away the old hierarchical society of the 1950s, with its narrow conception of status, and replacing it with an increasingly pluralistic and diverse culture. The deeply entrenched idea that status is a fixed resource, and striving for it a zero-sum contest, turns out to be false. The diversifying, anti-hierarchical forces of cool consumption supply new status. For this reason, we'll suggest that the proliferation of consumer lifestyles over the last fifty years is best seen as a solution to the Status Dilemma.

By the 1990s, the social changes that rebel cool had unleashed were giving way to a new kind of oppositional consumption. We refer to this as DotCool, in part because the norms it came to embrace are those that are valued in today's postindustrial society, variously referred to as a knowledge, information, or learning society. The last three decades have witnessed an explosion of new routes to status (labeled subcultures, lifestyles, consumer microcultures, consumer tribes, or brand communities) in our increasingly fragmented and pluralistic society. DotCool continues to branch out influentially today in the forms of hip consumerism, including ethical consumerism, political consumerism, and green consumerism. Indeed, although it may sound paradoxical at first because of the long moralistic shadow that's been cast over consumerism, responding to the challenges of climate change may depend on tapping into the human instincts that drive consumerism in ways we've seldom ever considered. But to see that—and to see how cool drives our economy and shapes our world—we need to explore the revolutionary science of how our brain creates us, the consumer.

2

THE THREE CONSUMERS WITHIN

It's early evening, and you've stopped at the grocery store to pick up a few items for dinner on your way home. You grab a cart and start strolling the aisles, quickly scanning the shelves up and down, left and right, tossing items in your cart as you go. You head to the checkout, casually glancing at the tabloid covers as you pick up a last-minute item; you exchange a few pleasantries with the cashier, swipe a card, fill your bags, and head on your way. Now, suppose we stopped you on the way out to ask you some questions about your shopping experience. Would you be able to remember the details? When pressed to recall how many brands of laundry detergent or cookies were for sale, for example, how accurately could you respond? Pressed to recall details like these, you might reply that you weren't really paying attention. In fact, you might add that you can't answer these questions because you were carrying on a conversation with someone, talking or texting on your phone, or trying to keep your child from tossing Pop Tarts into your cart. Few things seem more mundane and routine than grocery shopping. But wait.

The first issue worth pointing out is the sheer complexity of the decision challenges the supermarket creates for you and your brain. While the human brain evolved under conditions of often extreme scarcity, there is no better example of overflowing abundance than the typical supermarket, which contains thirty thousand or more items. Among these, many are virtually indistinguishable from one another, providing little or no objective basis for choosing one. There are nearly one hundred different varieties of laundry detergent alone, requiring you to choose among such fragrances as Apple Mango Tango, White Lilac, Vanilla and Lavender, Renewing Rain, Spring Renewal, Tropical Passion, Linen and Lilies, Meadows and Rain, Mountain Breeze, Mountain Rain, and Mountain Spring. As those fragrance names suggest, many of the thirty thousand items available to you have been shaped, tweaked, and fine-tuned by market surveys and painstaking focus groups—who apparently have a thing for rainy mountains.

From their chemistry to their names to the smallest details of their package design, the items confronting you at the supermarket have all been designed with the goal of giving them a competitive edge among the sea of alternatives before you. Not only that, but consultants with expertise in supermarket layout, so-called choice architects, have spent countless hours working out how and where those products should be displayed: positioning the bakery or the florist near the entrance so their fragrances will act on your olfactory brain and unconsciously prime you to spend elsewhere in the store; fine-tuning the details of the layout to change with the weather, the season, the day of the week, special events, and holidays.

How, then, are we able—in perhaps the most complex decision environment we've ever designed—to navigate and make decisions without paying much attention? To press the point a bit more, if you weren't paying attention, who—or *what*—was doing the deciding? Throughout the history of Western thought, in response to this question, people have pointed to a unique center of individual identity—a *me* or a *you* around which our behavior coherently revolves. We refer to it by different names: in religious contexts as a soul, in contemporary secular and psychological contexts as a self, in legal and medical contexts as a person, and in economic contexts as an agent. All these names point to something contained within you that endures over time, makes you the same person you were years ago, and allows you to connect events, days, and years together in a personal narrative—an enduring identity. We are so morbidly afraid of mental disorders and diseases, ranging from schizophrenia to

Alzheimer's, not only because they attack the body, but also because they attack the self. Imagine the personal terror you'd experience if your tastes, attitudes, personal traits, and desires changed from day to day, if you could not connect your thoughts and feelings right now to a personal past or an imagined future—if you only saw strangers when you looked through your old photos. Although the details are subject to considerable controversy, there is something deeply intuitive at the core of this idea of an enduring *you*. And it is intimately connected to the idea that what guides our actions is an underlying self, an identity whose uniqueness lies in a specific bundle of traits, tastes, aptitudes, and goals.

The concept of the unified self also plays a fundamental role in economic decision making and consumer theory. We think of a consumer as someone who has a set of preferences, a so-called utility function that ranks tastes. You are your preferences, in other words. The unified self can also project itself into imagined futures: our present self can plan for our future self, however imperfectly, choosing to save for retirement or to forgo a dessert today to look good at the beach next summer. Trading off a reward for our present self in favor of a reward for our future self is only sensible if we can conceive of ourselves as relatively permanent, with clusters of stable personality traits, tastes, skills, and so on. Otherwise, we would be planning for someone else's future. Indeed, many theories of addiction describe the addict as blind to the future—someone who no longer cares for his future self. A self trapped in the present.

The emerging science of neuroeconomics shatters the conception of the unified self. There is no coherent *you* behind your decisions and actions. The existence of a unified self as a central decision maker is an illusion, a convenient fiction, a largely unconscious rationalization. Combining neuroscience, computer science, and economics, the neuroeconomics perspective on how we make decisions, which we present here, views you as a confederacy of three decision makers, or "controllers"—each a kind of "pleasure machine" lurking inside you. The three pleasure machines are the Survival pleasure machine, the Habit pleasure machine, and the Goal pleasure machine.[1] Different evolutionary pressures have shaped each one. Each deploys its own strategies, values, information, and emotions to guide our behavior. And each plays by its own rules, typically below the level of awareness. And, as we'll discover, these pleasure machines were in place within humans to guide our decision making and behavior long before the rise of modern consumption, making modern consumption a reflection of the ancient purposes and strategies established within us long ago.

Indeed, the Survival and Habit pleasure machines are so basic to survival that they are shared in essential respects by nearly all sentient creatures. Think of your Survival pleasure machine as a set of evolutionarily important instinctive behaviors or reflexes, ones that are pleasurable because they are directly linked to survival, such as when you reach for food without even thinking about it—not so different from a dog's Survival instinct to happily eat up a table scrap he discovers on the floor. Think of the Habit pleasure machine within you as the one that underlies your habitual behavior, like that comforting morning cup of coffee with your newspaper. Think of the Goal pleasure machine within you as the one that makes to-do lists. It gives you the ability to weigh the pros and cons of various options in a deliberative, rational manner. In its basics, the Goal pleasure machine is probably common to all mammalian brains, and perhaps to some other animal species, but it has been reshaped by the unique demands of human social life to create social desires, values, and motivations that underlie much of our uniquely human economic life. Together, these three form what we refer to as a neurodynamic mind.

These pleasure machines often conflict with one another, and this conflict is at the root of many problems; it's why we often overspend and undersave, purchase on impulse, eat everything on the plate even when we know it undermines our dieting goals, and sometimes want things out of all proportion to the pleasure we'll receive from owning them. Within us, then, there is no centralized command, no charioteer holding the reins to steer us on a steady course, as Plato imagined. The reality is closer to a chariot race, each controller racing against the others, one momentarily taking the lead to gain control of our behavior, then another inching in front. Their constant race goes on largely below the level of our awareness; our behavior seems to us to flow seamlessly from a single self, while it actually shifts from moment to moment from one pleasure machine to another. In fact, the truth may be even more disquieting. The perception that our decisions stem from a unified self, a central decision maker, may be a bias that is built into us, a simple but misleading way to understand our own actions that hides their far deeper reality. If we really want to understand how and why we act, we need to abandon the idea of the unified self and its centralized command. We need to understand our consuming behavior as the dynamic interplay among the three pleasure machines within each of us.

The neurodynamic mind will be our Rosetta stone for understanding consumption—the foundation of a new science of desire and pleasure. By decoding the activity of the three pleasure machines, we'll see that there are three consumers within us: an instinctual consumer, a habitual consumer, and a social consumer.[2] Each gives rise to different facets of our consumption, making consumption a complex blend of often differing ends. This view radically changes our understanding of why we consume and provides the foundation needed to understand oppositional cool consumption.

Using this framework, we'll discover that cool is a social value rooted in impulses from the Survival pleasure machine and the Goal pleasure machine. Your Survival pleasure machine contains a "stick it to The Man" reflex that gets triggered whenever it senses someone or something is trying to hold you down. You didn't learn it. Your dislike of authority is built-in. It's a rebel instinct. In fact, you share it with your ape cousins, who also rebel every now and then by staging coups. Just listen to your favorite rebel anthem. Does it make you feel like you're not going to take it anymore?[3] That's because rebel music is like a drug delivered straight to your Survival pleasure machine, whose impulses are triggered by emotion. And it's why every anti-establishment movement has its own music. This is nothing new. After all, the 1830 Belgian Revolution started in an opera house. Your Goal pleasure machine, on the other hand, scours for things that might boost your status. Put them together and you get something that might boost your status as a rebel. That's cool in a nutshell.

To understand how our brain's pleasure machines work together to create oppositional consumption and how, as a result, cool became a central economic force, we first need to explore how the

neurodynamic mind gives rise to our consumption behavior far beyond the satisfaction of simple needs. We also need to put the three
pleasure machines in their evolutionary context to understand the
forces that shaped them long before they shaped contemporary cool
consumers.

Freud Redux?

Perhaps you're asking yourself if there is some evolutionary significance to having three different "you's" pulling your strings. That the
mind has multiple parts is, of course, not a new view. In fact, we
often consider Plato's theory of a three-part mind more than twenty-
four hundred years ago to be the origin of the Western conception
of the mind.[4] In addition, Plato's division between the rational, the
spirited, and the appetitive parts of the mind bears some similarity
to Freud's superego, ego, and id. Freud's view, like many theories
today, carves up the mind into rational and emotional faculties as a
fundamental and basic framework for understanding human behavior. Rooted at least as far back as Socratic thought, this reason-
emotion split continues to permeate our understanding of the mind,
as well as our consumer-behavior studies and marketing research.
Freud used the term "psychodynamics" to characterize how these
components interact to create our mental life and behavior. He was
especially interested in how their conflict created mental illness. It's
worth pausing here to ponder the similarities and differences between our view and Freud's theory, since we're going to return to
Freud throughout this book to highlight ways his influence led social critics down blind alleys to badly distorted diagnoses.[5]

Freud used the word "id" for basic, instinctual drives operating
unconsciously according to "the pleasure principle"—the biological
tendency to move away from the unpleasant tension of unfulfilled
needs toward the pleasure of getting gratification. Freud thought
the id's libidinous energy gave rise to early behaviors of infants and
children that drove their psychological development. Most notori-
ously, Freud believed that infants were sexual, since they were born
equipped with the id's drives. Hence, human development was a

psychosexual journey through various stages that focused on different erogenous zones, from the infant's "oral fixations" to the toddler's "anal" preoccupation with toilet training to the awakening of the small child's desire to "genitally" possess the opposite-sex parent. The repression of that forbidden and dangerous wish led to the development of the superego's puritanical conscience. The poor ego was stuck trying to balance the id's ravenous desires against the superego's punitive guilt. According to Freud, socially acceptable behavior depends on "sublimating" your id's impulses to rape and kill. Your creativity at work, for example, is just the redirected sexual energy of your id.

While this may sound melodramatic to us today, Freud's idea that we start life with a small repertoire of inborn behaviors that then expands over the course of our development is a sensible one.[6] In fact, we believe the Survival, Habit, and Goal pleasure machines follow a similar developmental scheme. Among primates, the human brain is the least mature at birth, weighing only about a quarter of its mature weight.[7] Not only does most of its development take place when social and environmental factors can exert their influence, but different regions develop at different rates.[8] In particular, the Goal pleasure machine has an extremely protracted development, lasting into the third decade of life.[9] This protracted development is one reason why neuroscientists increasingly regard teenage decision making as less mature than a young adult's. Indeed, neuroscientific evidence of the incomplete development of the teenage brain played a role in a 2005 Supreme Court decision ruling that capital punishment for minors was unconstitutional because their Goal system was still maturing.[10] This protracted process of development, with its dependence on social and cultural interaction, is one of the cornerstones of cultural biology, our view of how the developing brain and culture interact to create our mind.[11]

There may be some surface similarities between Freud's psychodynamic mind and our neurodynamic mind, but the differences are major. One of the most fundamental differences between the two is that Freud regarded the id as conflicting with social life. The id, in his view, has to be coerced into civilized life because its primary drives are sexual and aggressive. This taming of the beast is the job of the superego's guilt-inducing conscience. Considering Freud's speculation that society had its origin in an Oedipal drama—brothers banding together to kill their father, which induced an overriding sense of guilt that allowed civilization to form—it's not surprising that he would see sociality as inherently frustrating. Such perspectives, though, have largely been discredited in light of recent evolutionary insights. Indeed, it's the Survival system that likely houses our most basic social behaviors and many of the deepest social emotions that are the primordial glue of our social life. For example, the smell of a newborn baby activates the Survival pleasure system in new mothers more strongly than it does in women who have never had a child, helping to establish the bond between parent and child. The id as an asocial, sexual, and aggressive drive is among Freud's most pernicious mistakes.

There is no single decision maker that works well in the complex environments we inhabit. So the brain's design of decentralized control is a way of spreading the risk by having different decision makers for different contexts. The Survival system is fast and computationally cheap, but it is inflexible and involves only a relatively small set of stereotypical behaviors. The Goal system, in contrast, is flexible, but uses a lot of cognitive resources. The Habit system learns slowly, but is flexible and doesn't require a lot of cognitive resources. You probably don't have an *instinct* for tennis, but it's not ideal to use the Goal system to deliberate about every shot in a game, although you likely start out that way. The Habit system is just right: over time, and with practice, you build a skill and it requires very little in the way of cognitive resources. At some point in learning a skill like tennis, the Habit system starts to perform better than the Goal system. These systems needn't be in conflict with one another. Their interaction may be cooperative, and behavioral control may go to the system that is most likely to do best with the problem at hand.

Beyond Reason vs. Emotion

Traditionally, the distinction between our conscious and unconscious mental life has aligned with the distinction between reason and emotion. We typically regard an emotional decision as one that simply comes to us without deliberation. Consider, for example, the following scenario from the social psychologist Jonathan Haidt, which was designed to elicit an emotional decision: The neighbors across the street have a family dog that they care for deeply. One day the dog gets loose and is struck and killed by a car. The family decides to cook and eat the dog. Is their action wrong?[12]

If you are like most people, you had a strong emotionally aversive response to this vignette, and an answer to "Is it wrong?" just popped up without the need to think about it. According to a "dual systems" view of decision making, emotional systems underlie unconscious and rapid decisions, alternatively referred to as hunches, intuitions, or gut feelings. It's possible that despite your emotional reaction, you still deliberated about whether this unusual family's action was wrong or not. Perhaps you considered whether the prohibition against eating a dog was simply a cultural norm, a convention, and you overruled your emotions to conclude that the action was perfectly fine indeed. This is the prototypical way we think reason operates. It is conscious and deliberative, slow and thoughtful.

As basic as the reason-emotion dichotomy is to understanding ourselves, however, the neurodynamic view forces us to abandon it. The neurodynamic view regards emotions as the way the brain encodes the values each system requires to choose appropriate actions. The Goal pleasure machine, for example, uses emotions to represent value, so separating emotion and reason into separate faculties doesn't work very well. Let's look deeper into the three consumers within, how they animate our behavior, how they drive our consumption, and how they ultimately provide the keys to understanding the rise of the cool consumption that figures so centrally in our economy and our lives.

A Brief History of Happiness, Pleasure, and Desire

Before considering why we consume, we need to ask: Why do we do *anything*? So let's start at the beginning.

Aristotle's view of practical reasoning was one of the first systematic replies to this question, and it is his view that most of us implicitly assume when we explain our behavior. We act because we both desire and reason. We seek some things, he noted, because they lead to other things. For example, few people *desire* to sit through Human Resources seminars on workplace conduct. Attending HR seminars is what we'd call an instrumental desire. Indeed, according to countless surveys, many people dislike their jobs, and so desire them only *instrumentally*—for the income they provide. Again, many people regard money as not having any intrinsic value but desire it for the goods and services for which it can be traded.

Aristotle noted that this regress has to stop somewhere. Something must be valued for its own sake. What's at the end of this chain of desires? What do we want for itself and not because it will lead to something else? Aristotle's answer would have profound implications. He suggested that the ultimate end of human action was happiness—everything we do is ultimately in the pursuit of it. Happiness is desired for its own sake. It is the central purpose of human life. Every object of desire, then, is desired to the extent that it may bring happiness. Everything else is merely instrumental, a link in the chain of desires anchored to happiness.

Desire. Reason. Happiness. These are the key ingredients of nearly every analysis of human motivation and decision making since Aristotle. So deep is the link between desire and happiness that its breakdown is diagnostic of mood disorders, such as depression, and personality disorders, such as masochism. Indeed, as we noted in the previous chapter, consumerism is based on the belief that consumption and happiness are deeply linked. To understand why we consume, then, we need to understand our desires, our wants and motivations, and how (or whether) their satisfaction leads to happiness.

Let's start by considering happiness. Aristotle thought of happiness as human flourishing, along the lines of what we might think of as human potential today.[13] By the eighteenth century, however, happiness had become equated with hedonism and the idea that our behavior was dictated by pursuing pleasure and avoiding pain. This was nowhere more pronounced than in the British philosopher

Jeremy Bentham's 1789 statement, "Nature has placed mankind under the governance of two sovereign masters, pain and pleasure. It is for them alone to point out what we ought to do, as well as to determine what we shall do."[14] According to Bentham, it is a fact of our nature that everything we do ultimately reduces to calculations of pain and pleasure. But Bentham took the idea a radical step further: he suggested that the pleasure an action produces is the sole measure of its value. In other words, an action is morally good in proportion to the pleasure it produces and/or the pain it avoids. Bentham calls this good "utility."[15] It would be his protégé, John Stuart Mill, who would go on to formulate the utilitarian doctrine, whereby the moral value of every action was reducible to its utility. For both Bentham and Mill, utility was largely synonymous with happiness and pleasure.

Bentham's theorizing about utility contained one more crucial element: the belief that it could be objectively measured. If the moral value of an action was the pleasure it produced, and if this pleasure could be measured, then it was possible to create a science of ethics as well as a scientific method for creating laws and social policy. This was a radically important prospect that fueled the emergence of neoclassical economics, consumer theory—and neuroeconomics. Bentham in fact sketched some quantitative aspects of pain and pleasure, such as their intensity and duration, and outlined a proposal for using these considerations to make legislation and social policies more rational. But it would be another century before economists quantified human pleasure and happiness.

For this part of the account, we turn to Francis Ysidro Edgeworth. Photographs of Edgeworth reveal a quintessentially Victorian scientist from an Anglo-Irish aristocratic family of some fame, his middle name alluding to his Spanish mother's family. His carefully manicured beard suggests the pedantic and eccentric qualities he was known for during his three-decade-long reign as professor of political economy at Oxford, an eccentricity heightened by what friends described as a curious blend of British and Spanish gentlemanly habits.

Reputed to have read Homer as a child while perched high up in a heron's nest, Edgeworth was one of the most prominent British economists of his day and an influential figure in the development of neoclassical economics. He is important to our account for a device he imagined in a rather remarkable and curious 1881 book, *Mathematical Psychics* (no, the title doesn't refer to clairvoyants with a penchant for math). Interspersing fanciful quotes from the ancients, reflecting his formal education in the classics, with mathematical notation from the calculus of variations, Edgeworth sought to fulfill Bentham's project—to provide a fully mathematical account of human pleasure and happiness, a kind of science of pleasure (in his words, "the application of mathematics to the world of soul"). In so doing, Edgeworth imagined a device he called a hedonimeter. Here is his account of this device's workings:

> Let there be granted to the science of pleasure what is granted to the science of energy; to imagine an ideally perfect instrument, a psychophysical machine, continually registering the height of pleasure experienced by an individual, exactly according to the verdict of consciousness . . . From moment to moment the hedonimeter varies; the delicate index now flickering with the flutter of the passions, now steadied by intellectual activity, low sunk whole hours in the neighborhood of zero, or momentarily springing up towards infinity.[16]

Was Edgeworth's pleasure-measuring hedonimeter merely eccentric Victorian musing? Arguably, had this device been built in Edgeworth's time, it might have been as revolutionary as some of the other technologies that shaped the twentieth century. Edgeworth's attempts to create a science of pleasure were part of an ongoing debate among his contemporaries regarding how the emerging science of economics was going to define and measure its central idea, utility. A theory of value had always been at the core of economics, since the issue of what makes a commodity have value, and how that value is determined, is foundational to economics. Classical economists such as Adam Smith and David Ricardo had looked to the costs involved in the production of a commodity as the determinants of its value. Such labor theories of value hold that the value of a good is an intrinsic, or objective, property of that good, its "natural price." Economic value was, in short, an objective property out in the world.

But what if economic value was something different? Specifically, what if it wasn't an objective property of the world, but something subjective? Philosophers had long distinguished between primary and secondary qualities, where primary qualities were objective properties of the world and secondary qualities were the subjective effects objects had on our sensations. For example, wavelength is a *primary* quality of light. Color, on the other hand, is a *secondary* quality. Color is not an intrinsic property of objects. Objects have certain surface properties that result in light reflecting off of them at certain wavelengths, but color exists only in the minds of observers, whose brains have converted these wavelengths into a subjective experience. (The proverbial sound, or not, of the falling tree in a forest is another example.)

Classical economists had assumed that economic value was a primary quality—that it existed in the world and could be objectively measured by considering production costs. In the span of three years, between 1871 and 1874, three economists, William Stanley Jevons, Carl Menger, and Léon Walras, in England, Austria, and France, respectively, published revolutionary works all rejecting the classical theory of objective economic value.[17] Instead, they argued that economic value is subjective. According to Jevons, for example, it was wrong to say absolutely that some object has utility: the difference between an object and a commodity to Jevons was that the latter is valued by an individual because it affords pleasure or wards off pain. Our desires create utility, making economics for Jevons a science concerned with the "laws of human want." Indeed, according to these laws of human want, it is wrong to say that every equal part of some commodity has equal utility.

Jevons illustrated this with an intuitive food example, which we present with some updating thanks to an experience at one of Mario Batali's restaurants. Suppose you've reserved a table at one of the celebrity chef's restaurants, where you've looked forward to eating for a long time. You decide on the seven-course tasting menu, assuming each course will be an epicurean delight. The first course is sumptuous. The second course is lovely. The third course is delightful.

By the fourth course, you're starting to feel full. By the fifth course, you're starting to ponder whether there's an inconspicuous way to unbutton your pants. The thought that there are two more courses begins to fill you with dread. Of course, the final three courses aren't bad food, but their utility for you just isn't the same as the utility of the first four courses. This is simply an illustration, Jevons argues, of a law of human want: each increment of a commodity produces less pleasure than the previous one.

If we were to divide your meal into even smaller increments, we would produce the law of the variation of utility, or diminishing marginal utility, which would figure centrally in the rise of modern economics. This way of thinking about utility would allow economics to quantify the pleasure you received from the meal as a function of the quantity consumed. Your utility was maximized when the quantity of food consumed increased to the point where your final increment of utility, what would later be dubbed marginal utility, was equal to zero. Think of it as the total pleasure or happiness you got from the meal. It was this sense of marginal utility that would become the cornerstone of a theory of exchange and prices.

With these works, economics was coming to regard value as a subjective quantity, akin to color—a subjective human construction. This new way of thinking about value allowed economists to convert pleasure and happiness into a mathematical framework—although happiness might be the last thing most people associate with calculus. But the new view also led to a potential crisis: how to measure the actual utility consumers received from commodities. Since utility was thought of in terms of pleasure, economists would have to develop a method to measure pleasure if the new economics was to emerge as a science. Edgeworth was part of a contingent that believed utility could be measured directly in terms of the subjective experience of pleasure. He took his inspiration for this from an emerging scientific approach to psychology known as psychophysics. In the 1830s, Ernst Weber, a professor of physiology at the University of Leipzig, in Germany, had been working on how we experience changes in the world, such as a light that slowly gets brighter. By the 1850s, his student Gustav Fechner had developed a law that would be fundamental to the emergence of psychology as a science, relating what he described as just noticeable differences—the smallest change one could experience—to physical changes in the stimuli, such as slowly increasing a light's brightness, increasing the volume of a sound, or even adding weights to an outstretched hand.

In all these cases, Fechner found that the amount of change needed for a barely noticeable difference increases with the intensity of the stimulus, whichever sense is in play. When a light is very dim or a sound very soft, for example, a small physical change in its intensity results in a perceptible difference, but as the light gets brighter, or the sound louder, greater and greater physical changes in its intensity are required to make a perceptible difference. Edgeworth noticed that Fechner's graphs looked a lot like ones economists used to plot marginal utility. This inspired him to think that psychophysics would provide a scientific basis for utility.

Despite Edgeworth's enthusiastic anticipation that utility might be understood in the new language of scientific psychology, others, including Jevons, were less sanguine about the prospects of measuring utility directly. After all, though psychophysics might actually be a promising enterprise, the hedonimeter remained a distant hope, and there was little concrete success at the time in measuring pleasure directly. Those who were skeptical of a physiological foundation for utility looked for indirect ways to measure it. The most promising approach was rooted in behavior—it was observable and could be readily measured.[18] The behavioral approach to utility won the day, and economists eventually abandoned the idea that utility could be defined as the subjective experience of pleasure.

The Brokers and the Bees

Economics abandoned the connection between utility and its physiological underpinnings. But the quest to understand the physical basis of pleasure continued in other sciences, particularly behavioral neuroscience. Does looking inside the brain provide new insights into these age-old questions? One of the most famous neuroscience experiments dealing with pleasure took place in the early 1950s at McGill University. James Olds and Peter Milner implanted electrodes in various regions of rats' brains and allowed the rats to directly stimulate these regions by pressing a lever. When the electrodes were implanted in a region deep inside the brain containing the nucleus accumbens, the rats would obsessively press the lever up to two thousand times per hour for twenty-four consecutive hours. Hungry rats would actually choose to press the lever instead of receiving food. They even ignored potential mates in favor of pressing the lever.[19] It seemed Olds and Milner had discovered the brain's pleasure center.

In 1956, Olds wrote a striking popular article, called "Pleasure Centers in the Brain," in which he declared that the brain contains specific centers which, when stimulated, provide a potent reward that an animal will seek to repeat.[20] Olds thought that the rat's pleasurable experience was what made the behavior rewarding. As studies continued, scientists found that the brain circuits underlying these pleasure-seeking behaviors involved the chemical dopamine. By the 1970s, the "anhedonia hypothesis" of dopamine function emerged, stating that dopamine was a pleasure neurotransmitter and its release gave rise to the subjective experience of pleasure.[21] In a remarkable confluence of evidence, a great deal of work on addiction research likewise found that most drugs of addiction, from nicotine to cocaine, affect the brain's dopaminergic system. The anhedonia hypothesis is one of the most compelling explanations of the physiological basis of pleasure. Today, accounts of dopamine as the pleasure chemical are ubiquitous, from love advice columns to diet articles on cupcake cravings.

As compelling as this hypothesis is, it appears to be incorrect— an insight that played a critical role in the rise of neuroeconomics. One of the early clues that the dopamine hedonia hypothesis was incorrect came in the early 1990s, when Wolfram Schultz and his colleagues probed the function of dopamine neurons in monkeys during a variety of learning tasks.[22] They found that before learning, dopamine neurons are most active following the delivery of a reward, such as a bit of fruit juice. So far, this finding was consistent with the hedonia theory. However, when the monkeys learned that a tone preceding the juice reliably predicted the reward, their dopamine neurons were no longer active after the juice was delivered. Instead, the dopamine neurons were active right after the tone. Why didn't dopamine neurons fire after the juice was delivered?

The monkeys still seemed to enjoy the treat. There was another twist. When the researchers omitted the juice after the monkey had come to expect it, the activity in dopamine neurons went silent right at the moment when the reward was expected. Read Montague and Peter Dayan, postdocs along with Steve at the time in Terry Sejnowski's lab at the Salk Institute, recognized that dopamine neurons were doing something a lot more intriguing than just inducing pleasure.[23] The neurons were learning by making *predictions* about rewards. This is known as reinforcement learning, a ubiquitous form of learning in humans and other animals. Broadly, it occurs whenever we learn through feedback from the environment in the form of rewards and punishment. We learn in this way when our dopamine system makes a prediction that then isn't matched by our experience. To learn from such a mismatch, our dopamine system has to keep track of both when it's doing better than it expected and when it's doing worse than expected—both are opportunities to improve performance. As our neurons use this information to guide them, we learn to associate reward values with actions. In this way, our dopamine systems link learning with decision making. An animal can learn how to make decisions that lead to the most rewards.

Montague and Dayan realized that this is what dopamine appears to do, and they began exploring the possibility that it is involved in reward prediction, reinforcement learning, and decision making. This would become known as the brain's Habit system, which we refer to as the Habit pleasure machine to emphasize its link with reward. Unlike our goals, the Habit system learns to value actions— actions themselves, such as having a cup of coffee in the morning, become rewarding. Even if you have the goal of cutting down on coffee, your Habit pleasure machine will still seek out coffee because it values the *action* of having coffee more than the outcome itself. In 1995, the two published a study of the quintessential economic decision maker, *Bombus*—better known by its common name, the bumblebee. The worker bee's life is essentially devoted to a single task: gathering nectar and pollen for the colony. It is sterile, so it's not concerned about finding mates; it has few if any natural predators; and, unlike honeybees, it doesn't communicate to other bees in the colony about where to look for good flowers.

A bumblebee faces an extraordinarily complex set of challenges when it heads out of the hive in search of nectar. First, because a bumblebee doesn't have vast energy reserves, it needs to make its foraging as efficient as possible to maximize energy returns. Second, it has to compete with bees from other colonies and with other creatures for nectar, which is a scarce resource. To make matters more challenging, a bumblebee can't be certain where it will find nectar in its environment because the location of good sources is always changing. So the bumblebee must not only be capable of registering reward when it obtains something of value (the nectar), but it must also learn to predict rewards and to use those predictions to guide its foraging. Although economics is often thought of in terms of money, it's worth noting that *any* choice that is constrained is a form of economic decision making. The constraints can come in a variety of forms—in this case, the bee's need to balance the energy it uses to forage against the amount of nectar it finds.

When a bumblebee hovers over two different flowers, how does it decide which one to visit? To make a good decision, its brain must represent two critical values: the amount of expected reward (nectar) for each flower and the amount of risk associated with each flower. Risk is essentially a measure of how wrong its estimate of the amount of nectar in the flower might be. There's a striking fact about how the bee solves this problem that we want to draw your attention to. If it sounds at all like the problem you face when you're thinking about what stocks to buy, it's because the problems are oddly similar. In fact, a behavioral ecologist has used a Nobel Prize–winning economist's theory of stock portfolio selection to understand the foraging behavior of bees.

Before Montague and Dayan developed their computer model of bee foraging, behavioral ecologists such as Leslie Real at Emory University studied foraging behavior by creating artificial flower fields with preset levels of nectar. Real turned to the 1990 Nobel laureate Harry Markowitz's theory of portfolio management for individual investors to explain bees' foraging choices. Markowitz won the Nobel for his work on offsetting rewards and risks in the selection of stocks. A prudent broker, for example, is willing to forgo some potential reward to offset risk by choosing less volatile stocks (or less volatile combinations of stocks). More generally, most people are sensitive to risk—few people invest all their savings in lottery tickets or put everything down on a spin of the roulette wheel. Even though the potential payoff may be enormous, the possibility of losing it all stops most of us from making such risky decisions. Thinking about risk may seem like a quintessentially human prospect, but animals also have to consider the riskiness of their decisions. A bumblebee, like a prudent broker, will choose the sure and steady flowers over the riskier ones.

Another critical clue to how the bee forages came in 1993, when the neurophysiologist Martin Hammer identified a neuron in the bee brain that appeared to be involved in learning. "VUMmx1" uses a chemical called octopamine, which is similar to dopamine. Applying their computer model of dopamine reward learning to VUMmx1, Montague and Dayan could replicate bee foraging behaviors that Real had observed in real bees. In particular, they could provide a new understanding of how bees learn about the distribution of nectar among different-colored flowers in their environments and how to use this learning to guide their foraging.

This revisionary thinking about the role of dopamine in economic decision making provided a general theory for how creatures as diverse as bumblebees and humans learn about value in their environments and use that information. It also provided a new way to investigate the powerful role the dopamine system plays in human behavior. As this system is implicated in most forms of addiction, the deeper insight into its workings promised to recast the problem of addiction from one based on pleasure to a much broader problem that involved learning and predictions of reward. More pragmatically, the new understanding also provided a quantitative framework for thinking about the function of dopamine in decision making, which provided a link between biology and the mathematical frameworks economists were accustomed to. In other words, it presented biology to economists in a way that was intuitively appealing to them.

Inside the Human Brain

Technological advances also played a role in the rise of neuroeconomics. The development of fMRI (functional magnetic resonance imaging) in the late 1990s was critically important for the emergence of human cognitive neuroscience—the fusion of psychology and neuroscience—and subsequently of neuroeconomics, which made fMRI a central research tool. Before fMRI, neuroscience mainly used animal models, as there were frustratingly few ways to probe the brain basis of human behavior. Much of what we knew about human brain function up to that point was based on how human neurological injuries altered psychological processes such as memory, learning, and behavior. While this work was critically valuable in helping to guide cognitive rehabilitation, it was difficult to apply takeaways from these studies to the general population. Though it too has limitations, fMRI was a boon to the study of the brain basis of human psychology and behavior because it was noninvasive. Participants in fMRI experiments were not exposed to any risk, such as radiation, as was the case with traditional brain scans.

Around 2000, research using fMRI highlighted the role of dopaminergic systems in human economic evaluation, inspired by the nonhuman studies we discussed earlier.[24] Meanwhile, economists were beginning to see the promise of cognitive neuroscience as a means to test—and revise—economic theory.[25] By 2005, a Society for Neuroeconomics was created, and a few years later we began a graduate program at Caltech to train students in this emerging discipline.

This explosion of neuroeconomic research has converged on a number of brain regions that have consistently been implicated in economic decision making. In particular, two regions are of importance to us: the dopamine system, which we've discussed a bit, and which is referred to generally as the basal ganglia, and a region of the frontal cortex known as the ventromedial prefrontal cortex (VMPFC). As an illustration, ten years after Montague and Dayan's seminal studies, experiments in Steve's lab with Peter Bossaerts, a professor of finance at Caltech, and Kerstin Preuschoff, a graduate student at the time, found that the subregions of the basal ganglia encode estimates of expected reward and risk when we face a decision involving chance.[26] This study used a very simple card gambling task because the goal was to isolate the most basic elements of decision making. The participants played for real money, and they had to draw two cards from a deck of ten. A player could make a bet for $1 that the second card he drew would be higher than the first card he drew. Suppose that the first card the player picked was a 2. That would provide him with enough information to calculate the odds for the gamble. The expected reward, or expected value, is a quantity that decision theorists had long associated with the value of the gamble, in this case 80 cents. The risk of this gamble is greatest when the player draws a 5 because it has the largest variance in return. We use this sense of risk to measure the volatility of a stock, where earnings could either be spread over a few close values (a less risky stock) or could be all over the place (a risky stock). These experiments revealed that the ventral striatum was calculating these quantities in the way that financial theory predicted. That is, it was calculating in the same way that a broker would when making good investment decisions.

We did a follow-up study to see whether these two basic elements of decision making, reward and risk, were integrated anywhere in the brain in the way that economists think of utility.[27] We found it in the VMPFC. The ventral striatum has been implicated in economic valuation across a range of tasks, and, together with the basal ganglia, is the center of the brain's valuation system. We can think of it as processing value signals for the Habit pleasure machine, whereas the VMPFC processes value signals for the Goal pleasure machine.

It's worth noting that although the participants in these experiments felt anticipation after they drew their first card, often they weren't consciously calculating the expected reward and risk of these gambles and weren't even consciously aware of them. This underlies a critically important aspect of neuroeconomics: often, our introspective awareness has only limited insight into how our brains go about representing and solving a problem. One of the seminal breakthroughs of neuroeconomics is our ability to measure value directly in terms of brain activity. By merely monitoring activity in regions of your brain, we can relate the strength of that activity to the strength of your desires and values—even when you aren't making any decisions or there aren't any decisions to be made. What's more, we can obtain these insights into your desires and values even when you aren't aware of them, or aren't even aware of the fact that your brain is evaluating objects in your environment. This is because your brain constantly makes implicit value judgments and scours your environment for objects of desire, motivating and shaping your preferences and behavior *beyond your own awareness*. In addition, it's remarkable where this phenomenon takes place in our brains. Our value circuits are some of the brain's most powerful, basic, and ancient structures. In fact, they are *so* ancient that we share some value circuits with insects. These circuits underlie all our desires and give rise to all we value, from the most basic values tied to survival to the most human of desires and values, including love, beauty, and morality.

How Many Utilities?

Edgeworth was interested in measuring the amount of pleasure, or utility, we receive as the result of a choice we make. This fits into a long line of utilitarian thinking, which views the hedonic consequences of an action as the measure of its value. Today, this is known as "experienced" utility. We can think of experienced utility as the pleasure you obtain from the consumption of a good—say, when you're enjoying a Snickers candy bar. But that's not the only kind of utility you need to get through your day. To guide decision making, you need to be able to calculate how much pleasure you think you'll get from that Snickers versus the Kit-Kat beside it, or maybe a banana instead. This leads to an important distinction between predicted and experienced utility. Presumably, you picked the Snickers because it had the greatest expected utility for you.

Neuroscientists often distinguish between these kinds of utility as the difference between wanting and liking. If, after a few bites of the Snickers, you realize it doesn't taste very good, then you wanted it more than you like it. Your brain should learn from the experience we call disappointment and adjust its future expectations of Snickers. As it turns out, we may not be as good at this as we think. We sometimes persist in wanting things more than we like them. The Harvard psychologist Daniel Gilbert has drawn attention to this phenomenon of "miswanting." If you've ever set your alarm for a predawn wake-up, only to curse it when it goes off and instead hit the snooze button as you ask yourself what you were thinking, it's because you were wrong about how much you wanted to get up for that run.

Today, we know far more about wanting than liking. In some ways, this is because wanting is more interesting, as it generates behavior and guides decisions. Liking is sort of dull from a psychological point of view. For example, addiction is probably more a disorder of wanting than of liking, and we'll see that the distinction between wanting and liking becomes even more glaring in the case of habits, since many habits continue after liking is long gone. Indeed, the hallmark of a "bad" habit is a behavior that continues despite struggles against it—wanting without liking.

Earlier, we suggested that humans are (multipart) pleasure machines, but it might be more fitting to think of ourselves as pleasure-*seeking* machines, since the pursuit of pleasure underlies so much of our behavior. We probably spend a lot more time pursuing pleasure than consuming it! All told, four sorts of utility figure prominently in our psychology. We've already met predicted utility and experienced utility. There's also *remembered* utility, which is the recollection of experienced utility. And then there's *decision* utility, which occurs right at the moment of choice.[28] Your brain needs all of these types of utility to navigate your world: predicted utility guides your choices; experienced utility is the pleasure of consuming a reward, which enables motivated behavior in the first place. Comparing experienced utility to predicted utility is a form of learning, which—at least in theory—should lead to better predictions in the future. And if we didn't have remembered utility, then we wouldn't be able to use our memories of pleasurable (or not) experiences to guide our decisions. When you're browsing a bookstore or a wine store, for example, you're constantly consulting your remembered utility for different authors or wines to guide your choices.

Remembered utility turns out to be a strange thing. It's not just a simple recall of, say, how pleasurable an experience was and how long it lasted. Instead, it depends on two things about experienced utility: its peak intensity and its intensity near its end. Daniel Kahneman refers to this as the peak-end rule. In a famous study of patients undergoing colonoscopies, he found that people's recollections

of the (un)pleasantness of the experience tended to neglect its duration.[29] Instead, their rating of remembered unpleasantness was
based on the most intense moment of pain and the pain during the
last few minutes of the procedure. Paradoxically, making the procedure longer by adding some minutes of mild pain at the end, rather
than stopping sooner at a higher level of pain, caused people actually to recall the entire experience as less aversive.

If remembered utility sounds strange, things get even stranger
when we realize there's no unified, single decision maker within us.
As we said at the beginning of this chapter, there are actually three
decision makers within us: the Survival, the Habit, and the Goal
pleasure machines. It's possible that each system has its own set of
utilities. That is, the Survival system has its own predicted utility,
decision utility, experienced utility, and remembered utility, as do
the other systems. If this is the case, then it would bring the total
number of utilities motivating each of us to a dizzying dozen. Work
in Steve's lab and elsewhere suggests that this may be the case.[30] So in
place of the single experienced utility, or pleasure as liking or enjoy-
ment, we need to add three additional types of utility.

The idea that we are a complex agglomeration of pleasure machines, each with many different kinds of utilities, is not without
precedent. In his defense of utilitarianism, John Stuart Mill argued
that there were at least two distinct types of experienced utility.
The proposals of Mill's mentor, Bentham, regarding pleasure
and happiness had been ridiculed by some critics because he argued that pleasure differed only in quantity, and that if someone
received as much pleasure from a game of push-pin as from poetry,
for example, then that would make push-pin as good as poetry.
Thomas Carlyle objected that this sounded like a "philosophy of
swine" since it made no distinction between a pig's pleasure from
rolling around in the mud and the more rarefied pleasures that
would befit a Victorian gentleman. Mill responded by adding quality to quantity as a measure of experienced utility in an attempt to
distinguish between higher and lower pleasures.

To put this in the language of the neurodynamic mind, is there a
difference between the Goal pleasure machine's experienced utility
and the Survival pleasure machine's experienced utility? Does the
experience of a fulfilled goal differ from the experience of a satis-

fied Survival desire? Does the remembered utility of each differ? To date, we know of no neurobiological studies that address these questions, although many central issues related to well-being hinge on them. For example, we may receive experienced utility from the satisfaction of a Survival desire, like eating two slices of pie, but then we have to live with a remembered disutility—regret—if it thwarts a goal of ours, like losing weight. What is particularly frustrating about these sorts of cases is that our Goal pleasure machine's remembered utility doesn't change our Survival desire for dessert.

These dynamics likely underlie the high failure rates of long-term goal-directed behaviors, such as diets, and the situation gets really strange when we add habits to the mix. The Habit pleasure machine works by valuing actions instead of their outcomes. In other words, if you habitually have a cup of coffee in the morning, it's because your brain values the action of having coffee—and you'd still go to the coffeemaker on autopilot tomorrow morning even if you had a bad cup of coffee today or were trying to kick your coffee habit. When you wake up in the morning, your Goal system doesn't need to predict how much pleasure you'll get from a cup of coffee. Chances are you didn't even think about whether you wanted a cup or not. The Habit system acts like a pleasure-guided autopilot. What this means is that habitual actions can retain value even after the outcome is no longer pleasurable. It also means that even if you formulate a goal to stop having coffee in the morning, the Habit system still drives your behavior. This is the essence of a habit. Your Goal pleasure machine may set the utility of a cup of coffee to zero, but the Habit pleasure machine still values the action and seems disturbingly able to override our goals.

Beyond Awareness

So our decision making is deeply complicated by the presence of multiple decision makers within us. There are as many as a dozen different utilities within us. This diversity is evolution's version of division of labor. Think of each value machine within you as a different division of a company. Each value system has a different job to do, and so deals with the world differently than the others by using its own sources of information and its own learning strategies. The Survival pleasure machine is like Shipping and Receiving, which works fast to get the job done. The Habit pleasure machine is like middle management, which uses standard practices to take care of the day-to-day business. The Goal pleasure machine is like the CEO, who plans for the company's future and gets called in if there's an emergency (and, like the Goal system, may be less responsible for the company's success than he or she thinks). Overall, having specialized divisions makes us better adapted to complex environments. That said, this division of decision-making labor also creates myriad opportunities for inaccurate utilities to emerge, through misremembering, through neglecting the ways in which our brain adapts to pleasure (responding with decreasing intensity to repeated stimuli), and through conflicts within the neurodynamic mind. And this picture is made even murkier by the fact that many of these processes operate below the level of our awareness—and that there may be good evolutionary reasons for giving us only limited and inaccurate insight into their workings.

We can get along in our complex world because our pleasure machines work below the surface of our awareness, scanning our environment for value and guiding our decisions and behavior. In fact, the truth is a bit more radical than that. What you are aware of is a tiny slice of what is out there. There is simply far too much information within the range of our senses for us to be fully aware of the world as it is. Luckily for your sanity, your brain is a supreme editor, keeping the raw footage of experience beyond your awareness. Were you aware of that raw footage, life would be a buzzing confusion of such magnitude that it would be a kind of madness beyond any surrealist's imagination. Sights and sounds would be disconnected and out of sync. Color, motion, tastes, and touch would all be helter-skelter, fragmented, disembodied, and jumbled. By comparison to the raw footage of our experience, even our dreams and the wildest surrealist fantasies are orderly and highly edited. Our conscious life—as rich and complex as it may seem—is only the final edit, the very tip of barely fathomable mysteries lying beneath. Most of the raw footage remains forever hidden to us, lost on the editing-room floor that is the brain's unconscious and secret life.

We used to think that the brain's job was to replicate every excruciating detail of the outside world, as a camera would, presenting us with a full and accurate picture of reality. We now know that this is not how the brain works. The brain is not in the business of delivering truth. After all, our senses exist to aid survival, not to serve some deeper philosophical pursuit of truth. It's sometimes better for the brain to guess the presence of a predator from brief, incomplete information than to wait for all the evidence. But, even more intriguing, recent evidence reveals that when we have the luxury of time. we still don't see the world as it is. Studies of replacement of one person or thing by another and of "inattentional blindness" (not seeing something in plain sight) show that we don't see the world as accurately as we believe.

In a stunning rebuke to the idea of the mind's eye as a camera, when you focus on one particular part of a video, a person can walk right through the scene in a gorilla suit and stop and wave, and you won't even notice. In another example, people often fail to notice when a person they are giving directions to is surreptitiously replaced by a different person (in one such study, workers carrying a sheet of plywood walk between the two and the switch is made). That proverbial frog may not notice slow changes in water temperature until it is too late, but we also fail to notice changes to a picture when the changes are gradual or happen out of our sight.

Our experience of the world, then, is profoundly colored by the mandates of survival, often delivering to us just enough highly selected information to keep us alive. For this reason, our brain's editor is no impartial spectator. It assembles as sensible and coherent a rendering of reality as we need to navigate our world. Why aren't you equally aware of your whole visual world? Why is the focus of attention so narrow?

Although your brain is a powerful computer, it has its important limits.[31] In particular, attention is a limited resource, and only a fraction of the information affecting your senses makes it into your awareness. Below your awareness, your brain is constantly monitoring what is invisible to your mind's eye. Indeed, as you read these words your unconscious brain might detect a sudden motion out in the periphery. If it does, it will command your eyes to shift, drawing your attention to whatever is moving out there. In other words, on the basis of information you are not aware of, your brain will make a decision about where your attention should be directed, without waiting for any input from you. The overwhelming majority of decisions your brain makes are of this kind: fast, largely involuntary, and intimately linked to value. You've probably had the experience of driving somewhere and suddenly realizing that a chunk of time has passed and you have no recollection of paying attention to the road or your progress. While that can be disconcerting, it is the brain's way of letting it take care of the driving while you daydream or cruise radio stations. Your brain may suddenly interrupt your reverie due to an impending threat, such as a stopped car ahead. It does so only because it is able to make an unconscious value-based decision: your unconscious brain has been monitoring your drive. Our unconscious brain organizes our conscious life around what is of value to us in our struggle to survive.

There are two implications here. First, what we become aware of is determined in part by our unconscious brain's search for what is of value to our survival—either potentially positive rewards or negative threats. We don't simply see the world as it is in a disinterested manner. Rather, our perception of reality itself is value-laden. Second, even if we don't become aware of something of value in our environment, often our unconscious brain has registered its value and may have created unconscious planning behaviors based around it.

In one intriguing fMRI experiment, Anita Tusche, Stefan Bode, and John-Dylan Haynes separated participants into two groups in order to test whether conscious awareness matters as much as we think it does.[32] While their brains were being scanned, the first group saw images of cars on a screen in front of them and were asked to rate how attractive the cars were. While their brains were being scanned, the second group performed a visually demanding task like the type your eye doctor gives you to test your peripheral vision. Meanwhile, images of cars flashed for a few seconds outside their focus of attention, so that subjects never became consciously aware of them. Later, the researchers showed the car pictures to both groups and asked them to state their willingness to purchase the cars—as if they were a focus group. Neither group knew beforehand that they would be asked about their purchasing intent. The experimenters then investigated whether there were patterns of brain activation during the first task that predicted the participants' later purchase choices—that is, whether there is a "buy" signal in the brain that forecasts hypothetical consumer choices of the sort marketing departments often look for in focus groups. This line of inquiry would seem plausible in the case of the participants who saw the car images front and center. In fact, the investigators found that the VMPFC encodes a "willingness to purchase" signal that activated when the group first saw the car images. This is interesting, since the participants weren't asked about their willingness to purchase the cars until after they completed the first part of the test. What the scientists determined was that the brain evaluates these sorts of stimuli in terms of potential purchases even in the *absence* of any specific instruction to do so.

But here's the kicker. The same "willingness to purchase" signals in the VMPFC were present in the participants who weren't even consciously aware of the cars during their visually demanding task. Their unconscious brain responded to the car images that were outside of awareness *and* made economic evaluations of them. The upshot is that awareness doesn't seem to be necessary for the brain to make even these complex evaluations. Consumer-related goals as complex as these can be activated unconsciously, and may guide our behavior and choices without our ever being consciously aware of it. These results overturn the canonical model of the "conscious deliberator." Because of these complexities, it's worthwhile to take a deeper look into the biology of desire and pleasure within the neurodynamic mind through a real-world example. Let's return to the desires closest to our survival, those related to foraging, or in the case of us modern humans, grocery shopping.

Zombies in the Supermarket

At the supermarket, your Goal pleasure machine is the one that has made a list at home and then deliberately seeks those items within your budget. It also problem-solves as you go along when you have to weigh the pros and cons of various options. Many of us just buy the same old salad dressing we always do, relying on habit. But if you've ever grocery shopped with someone who takes ten minutes to decide on what kind of salad dressing to get—muttering on and on about the pros and cons of various types and brands—you've seen the Goal pleasure machine in action. The Goal pleasure machine is your conscious deliberator. It's the closest to the standard model of economic choice, which supposes you have a single set of well-defined goals or preferences that you maximize rationally with your budget in mind.

How much of a trip to the supermarket is governed by the Goal pleasure machine? Well, consider the fact that two out of three items in the typical shopper's cart aren't planned purchases. In other words, most of your supermarket purchases are the product of your Survival and Habit pleasure machines. Our Survival pleasure machine is the most basic. We can think of it in loose terms as being programmed by evolution.[33] It contains values for things that are critical for survival and links them to a set of relatively inflexible but quick-to-launch actions. If you've ever chatted with a friend beside a plate of treats and found yourself suddenly holding a donut without realizing you were thinking about it, you've experienced your Survival pleasure machine at work. Or maybe you suddenly caught yourself involuntarily stooping due to a looming shadow, or maybe you've jumped, cowered, or run after hearing a loud bang before being fully aware of the sound and what caused it. These behaviors are basic to survival and flow from circuits deep in your brainstem that evolved long before humans.

Clever design has made the modern supermarket a veritable Survival wonderland, with smells, displays, and packaging all intended to elicit these semi-reflexive behaviors. What's more, the Survival values in your brainstem are modulated by the hypothalamus, which detects states of your body, such as hunger and thirst, and alters your Survival values accordingly. A striking example of this can be found in the shopping carts of dieters, which are likely to be filled with more calorically dense items and fewer fruits and vegetables than those of non-dieters—the very food choices that sabotage diets! Here's what's happened: The dieter's hypothalamus senses a caloric deficit. To the hypothalamus, that can mean only one thing: "You are starving!" The right course of action is to adjust the Survival value system, upping the value of calorically dense foods while lowering the value of calorically sparse foods. These altered values create cravings for foods such as potato chips, bread, pies, ice cream, and other calorie-rich fare. Without being aware of what the hypothalamus is up to, the dieter finds a cart full of the wrong foods.

The dieter's cart contains a number of important lessons. The Survival pleasure machine, along with the hypothalamus that adjusts its values, operates without consideration of the goals of other systems. In this case, it is completely unaware of the dieter's Goal value system—which may have a goal as abstract as knocking 30 points off a bad cholesterol level in six months. The ensuing conflict between the two pleasure machines is not one of reason versus emotions. Instead, it's one of differing timescales. The Survival pleasure machine doesn't know about goals and is mainly concerned with short-term issues, such as appropriate responses in our current environment. Of course, the result is that it all too frequently undermines our longer-term goals. A somewhat darker possibility is that when there is a clash between our Goal and Survival systems, evolution has rigged things to bias our behaviors toward the Survival system, especially in the context of two forms of behavior intimately connected to survival (our own and that of our genes) that the Survival system strongly regulates: food and sex. Why else would an otherwise shrewd politician risk a successful career for a brief tryst? Why else is obesity becoming a global pandemic despite entire industries dedicated to fighting it? While we are not enslaved by our Survival system, its pull is strong and its myopia is great, and we inevitably fall back into Survival patterns despite concerted efforts to avoid them.

There is now evidence that the Survival system can also have a dramatic effect on how much we are willing to pay for goods. Ben Bushong, a graduate student at Caltech at the time of the study, along with Lindsay King, a Caltech undergraduate student at the time of the study, and the economics professors Colin Camerer and Antonio Rangel, wanted to know if the ways in which a good was presented had an effect on how much consumers were willing to pay for it.[34] Specifically, they were interested in whether the physical presence of a good altered our willingness to pay, since the physical presence of the good would obviously trigger Survival processes known to neuroscientists and psychologists as "Pavlovian consummatory mechanisms"—preprogrammed responses such as the impulse to reach or salivation. They presented snack items to participants in three different forms, either text (like the words "potato chips"), a picture of the snack item, or the actual product placed in front of them. They found that people were willing to pay 40–60 percent more for an item when it was physically in front of them. They reasoned that this increase in willingness to pay was a side effect of the Survival system: the physical presence of a tasty stimulus triggers the Survival pleasure machine, which automatically launches the evolution-programmed response, including motor behaviors to contact the good. They called this the real-exposure effect.

Bushong and his colleagues added a few twists to their experiments, which also highlight the Survival system. First, they tested whether having participants viewing a picture of a snack and swallowing a small sample of it would increase their willingness to pay. They found that it didn't have an effect. This goes against the idea that experienced utility is responsible for the real-exposure effect. Second, they placed a Plexiglas barrier between participants and the snack items. They wanted to keep the sensory information the same, but introduce a physical obstacle to Survival responses. In fact, the mere presence of the transparent barrier reduced willingness to pay to the same levels elicited by the picture presentation. As the authors note, this has a number of real-world implications: forget about providing samples to customers, and if you want to sell more desserts at a restaurant, put them on a dessert tray and cart them to the tables. And don't cover them with a glass dome!

Modern supermarkets, of course, have incorporated many design elements that engage the Survival pleasure machine, many arrived at through trial and error, but some carefully crafted by "choice architects." Indeed, you've likely never given much thought to the design of a shopping cart, but changes to its design can have large effects on consumer decisions. In one study, for example, researchers found that putting a line across the width of shopping carts with a sign asking shoppers to place fruits and vegetables in front of the tape line and the rest of their groceries behind the line resulted in consumers increasing their fruit and vegetable purchases by 102 percent.[35]

The Survival pleasure machine determines many of our consumer choices. At the other end of the decision spectrum, the Goal system underlies the purchases we've planned for ahead of time and the purchases we deliberate about as we shop, perhaps trying to problem-solve the menu for an upcoming dinner party to satisfy the preferences of all our guests. In between these two is the Habit system, which manages much of our supermarket behavior—and our lives more generally. You may have had the experience of visiting a foreign country and shopping in a supermarket there. Chances are you felt lost doing so, in no small part because supermarkets differ so much across countries. The products were unfamiliar and the layout was different, making it impossible to rely on your past experience. In such a case, you needed to resort to your Goal system, and you likely had to devote considerable attention to your shopping. This is even true to a large extent when shopping in a different company's supermarket in a different town, even in your own country. This reveals that over time you can "overlearn" even an extraordinarily complex decision environment such as a supermarket. You become so familiar with the environment that performing the task becomes automatic, and you can do it without thinking about it.

Typically, this process is accomplished through repetition and practice, which provides sufficient experience for the Habit system to take control. "Automaticity," or the ability to do tasks without devoting conscious resources (that is, the Goal system) to them, is an important capacity that underlies many of our daily activities. To see how this shapes our consumption, we'll need to look into the Habit pleasure machine and its intimate connection to the object of its desire: the brand.

Brands on the Brain

Many people consider brands to be a relatively late invention of modern capitalist societies. This late-emergence view also underlies a growing anti-branding movement, epitomized by Naomi Klein's *No Logo*, in which Klein argues that brands are the problematic consequences of late capitalism. While this conveniently "problematizes" brands for critics, it's historically inaccurate. As the archaeologist David Wengrow reveals, commodity branding was evident as early as the fourth millennium B.C., thousands of years before the rise of consumerism. Branding emerged alongside the earliest large-scale economies.[36] Wengrow suggests that branding helped to solve problems of quality control, authenticity, and ownership that arise in any large-scale economy that mass-produces commodities. The brand was, in short, a signal that conveyed value for consumers. Prior to brands, Wengrow argues, it was difficult for consumers to determine the quality of goods before purchases because the quality and quantity were not standardized. Goods had to be examined and measured, and the flow of goods depended on networks of brokers in which trust was essential. This typically restricted trade networks to ones based on kin. These significant structural limitations on economies were solved in part with the invention of brands, which standardized commodities and allowed for expanded information flows among consumers and producers. So brands are not modern inventions at all.

As intriguing as Wengrow's findings are, we believe that brands are ubiquitous for an additional reason. Your Habit pleasure machine is primed for brands. A brand is its way of making sense of the world. To see how, let's look deeper into the natural affinities between brands and brains.

In 2004, Read Montague and his colleagues at Baylor College of Medicine performed a modified Pepsi Challenge study in combination with brain imaging.[37] It was one of the first studies of branding's impact on the brain, and it has become a classic in the field. The aim of the study was to test how brand knowledge affects our experience of drinking two soft drinks, Coke and Pepsi. Most of us are familiar with the "Pepsi Challenge" marketing campaign, which began in 1975, and involved a blind taste test of the two competing drinks. The fact that Pepsi often came out as the winner in this test made Montague and his colleagues wonder why Coke was the market leader if people appeared to like the taste of Pepsi better. The team's first step was to conduct a blind taste test to determine whether someone's brand loyalty aligned with their taste preferences. By re-creating the Pepsi Challenge, they wanted to confirm the results of the campaign we all saw on TV. Sure enough, although the participants were about evenly split in their stated preference for either Coke or Pepsi, the blind taste test results revealed that participants' stated loyalties were not significantly correlated with their blind taste preferences. Taste didn't seem to underlie brand loyalty.

The scientists repeated the blind taste test, this time with the participants placed in the scanner. When people got a sip of each soda, not knowing what brand it was, a region of their brain involved in subjective pleasure, the VMPFC, activated. The amount of this activation was greater for the sips people said they liked more. But, just as in the first blind taste test, the sip people said they liked more was often not of their preferred brand. At the level of the brain, then, brand loyalty, taste, and the taste's subjective pleasure didn't seem to match up. The answer to this puzzling mismatch came when people were scanned while tasting each soda—this time knowing its brand. When people sipped Coke knowing it was Coke, it activated brain regions underlying memory, emotion, and recollection. So the implication is that the emotional memories evoked by the Coke brand *modulated* the brand experience—a paradigmatic case of "emotional branding." In contrast, knowing that Pepsi was Pepsi had no effect on a participant's brain.

A subsequent Pepsi Challenge experiment provided even more evidence for emotional branding.[38] Researchers repeated the Pepsi Challenge, but this time included a group of people who had damage to their VMPFC alongside a group from the general population. VMPFC damage results in reduced emotion and emotional memory. As in standard Pepsi Challenges, both groups showed a preference for Pepsi in blind taste tests. Many of the people from the general population reversed their preference when they knew what brand it was, illustrating the behavioral switch underlying the "Pepsi Paradox." In contrast, the VMPFC participants didn't switch their preferences. These patients were "brand blind" because the effects of branding—building emotional associations—had little impact on them. Since emotional branding had little effect, these patients simply followed their taste.

The Pepsi Challenge series of experiments reveals how emotional branding alters brain responses. A 2007 experiment by Hilke Plassmann and her colleagues at Caltech shows an even more remarkable effect—how our expectations shape our experience. Their experiment was a sort of "in-the-scanner" wine tasting. People thought they were tasting four different wines, a $5, a $10, a $45, and a $90 wine. In reality, the supposedly $10 wine was really the $90 wine while the supposedly $45 wine was really the $5 wine. The question was, would people enjoy the $5 wine more when they thought it was a $45 wine and would they enjoy the $90 wine less when they thought it was a $10 wine?

Not only did what people said about their experiences agree with this prediction, but their VMPFC activity did as well. In other words, the expectation actually changed their experience. Expectation shapes experience. Although this study used price information, the implications are clear enough for brands. Brands carry information about quality, just as we saw in the case of the ancient brands. Had the wine-tasting study been conducted with known brands of wine, it likely would have resulted in similar findings. Brands shape our experiences; the cultural information a brand carries can contribute to economic value by creating more utility in the form of subjective pleasure. The converse is also true: a brand's poor image can actually damage the experience of its products. Even comments by an executive, such as the now-infamous remarks of Lululemon's Chip Wilson, can rub off on the product.[39]

If brands actually add value to the experience of a product by triggering strong emotional memories, how do brands get their value in the first place? The answer has important implications, in terms of not only how we learn the value of a brand, but also how we use brands to make decisions. Both insights stem from a fact we saw earlier: allowing the largely unconscious Habit system to guide our behavior facilitates our consumer decisions. In fact, letting our unconscious decide for us often results in not only easier but also better decisions.[40] We suspect that one reason why people don't eat more fruits and vegetables is because picking them out involves relatively hard decisions: unlike packaged products, they require us to deal with individual variability and nonstandard elements—for instance, how do you decide which melon is ripe? We don't know of any studies on whether this challenge is daunting enough to affect purchasing behavior, but the number of online articles just on how to choose a ripe mango suggests it's a problem. Recall that the essential element of the Habit pleasure machine is that it attaches value to actions rather than outcomes, allowing us to behave on the basis of stored action values rather than having to compute outcome values for every decision. And brands are like habits.

The basic brain processes involved in brand value formation were revealed in a study conducted by our Caltech colleague John O'Doherty.[41] O'Doherty asked participants to rate four different juices: blackcurrant, melon, grapefruit, and carrot. After this, participants were scanned while they tasted a sample of each juice. Before they got each juice, the subjects were briefly shown a symbol to associate with it. At the beginning of the experiment, the ventral striatum responded according to how much the participants liked the juice. As the experiment continued, however, this brain response (a higher or lower activity level depending on the amount of pleasure) transferred from the juice to its associated symbol. Each symbol began to work like a cue that predicted the amount of reward to be delivered—in other words, like a brand. If this sounds like the honeybees and the monkeys we met earlier, it's because brands tap into the same brain system in us.

Of course, brands are much more complex than the simple symbols used in O'Doherty's experiment, but there are a number of fundamental connections between the symbols and brands. The participants weren't told to learn which symbols went with which juices. They made no conscious effort to do so. Their unconscious brain was keeping track of these connections, and unconsciously learning. When we see a familiar logo, whether it's a Starbucks mermaid or the Golden Arches, our brains respond with an unconscious registering of the associated values we've come to build around those symbols as well. In fact, the way your Habit pleasure machine learns reveals the specific ways brands get into your brain. Your Habit pleasure machine learns when there are mismatches between what it predicts about a reward—like how good that cup of Starbucks coffee is going to taste—and the actual taste. When your predicted reward and delivered reward are the same, your value of that brand doesn't increase. The only way a brand's value can increase is when some brand experience exceeds your prediction. Of course, if your predicted reward is greater than what you actually experience—if the brand disappoints you—your Habit pleasure machine downgrades that brand. These negative experiences are much more important to your brain than positive ones. Have you ever gone back to a restaurant that gave you food poisoning? This presents significant challenges to brands, which are seeing a dramatic decline in loyalty in recent years, particularly among younger consumers.[42]

In fact, your brain harbors another reason why brand loyalty may be declining. Our brains face a basic conundrum that affects all foragers and shoppers, from bees to us. In science it's known as the exploration-exploitation trade-off; everywhere else, it's known as the "what have you done for me lately?/the grass is greener on the other side" problem. Consider that honeybee foraging on yellow flowers. If its environment isn't changing, then once it's found a reliable predictor of reward (the yellow flowers), it should continue visiting yellow flowers. The trouble is, its environment is changing all the time. By sticking with the yellow flower, the bee may be missing out on an even more rewarding blue flower. Every now and then, the bee's brain will decide to check out a blue flower, especially if the yellow flower is stuck at the same reward level. This maneuver is wired into our brains, and we suspect it has far-reaching implications for brands. For one thing, our brains treat a brand that's not improving (exceeding our expectations) as if it's getting worse. Also, our Habit pleasure machine will treat areas of rapid change, such as technology, as ones that need more exploring than exploiting. Increasingly small differences in quality and more-crowded markets only compound the challenges of building a brand into a habit.

Brands build emotional connections between our memories and our consumption. These emotions are critical to our consumption, but somewhere in our evolutionary past something changed in the way our brains connect emotions with consumption. New kinds of emotions emerged. These strange new *social* emotions let us connect products to our social identity, our sense of who we are—and of how others see us. Soon after, we possessed the ability to use *things* to communicate who we are to others. When that happened, a whole new kind of economic value emerged, one that would change the world.

3

COOL ON THE BRAIN

Looking through the control room's window, we can see the outline of Lisa's body as she lies deep inside the maw of Caltech's brain scanner. Images of cool and not-so-cool products and celebrities appear on a screen just above her eyes. In the noise of the everyday hustle and bustle of life, colored by profoundly complex social and commercial messages, hazy forces have worked on her to shape how she responds to these images. Just as the forces of nature operate on us in ways mostly beyond our awareness, so too do these social forces work mostly on our unconscious, shaping our beliefs, desires, and feelings in ways we seldom really understand or recognize. Somehow, Lisa's brain translates everything she associates with each image into a pattern of activity that represents her conscious and unconscious responses to that image. This pattern will include such things as Lisa's understanding of the instrumental function of an item (for example, the basic value of a watch is its accuracy in keeping time). But her brain's response will include more than that. Any watch these days can tell time with more accuracy than most people need. What is it beyond these instrumental functions that distinguishes some watches from others? What makes one an object of desire and another an object of indifference or of scorn and derision? What makes some products so desired that people will stand in line all night for the chance to purchase them, while other products will spell financial ruin for their makers? With this experiment we are going beyond the limits of introspection and the noise of everyday life to look directly into the brain to see how these forces work on us. For this reason, we've asked Lisa to do nothing more than simply experience the images as she would strolling down a city street peering into shop windows, or flipping through a magazine—we need no other responses because her brain's pattern of activity will tell us more about how these images affect her than will her own conscious reflection.

Lisa, our first participant in the study, is Lisa Ling, who at the time (2004) was the host of the *National Geographic Ultimate Explorer* series on MSNBC.[1] The experiment started as a query from Lisa—she was working on an episode called "CoolQuest" and wanted to know whether it was possible to measure cool in the brain. The question struck us as unusual at first. But the more we talked about it with friends, the more we realized that everyone could relate to cool. Everyone could recall at least one episode from their teenage years when their parents tried to make them wear some uncool hand-me-down or wouldn't let them get something the cool kids were wearing. Years later, the memories stuck and still stung. For Steve, it was an eighth-grade standoff in a shoe store, holding out for the Adidas the cool kids in high school wore instead of the North Stars middle school kids wore. North Stars had only two stripes. Adidas had three. How could one more stripe mean the difference between being cool and being exiled to the lunch table with the geeks and dorks? Even among scientists you hear the word "cool" all the time—and we're probably not hurting anyone's feelings when we say scientists don't exactly set the highest bar when it comes to cool. Here at Caltech, Richard Feynman is revered not just because he was a Nobel Prize–winning physicist, but also because he was pretty cool for a scientist.

The more we talked with people, the more we also realized that cool is something people *sort of* think they understand. But we kept hearing the same thing. Cool was like Supreme Court Justice Potter Stewart's famous definition of pornography.[2] You knew it when you saw it. But it wasn't something you could really *define*. In fact, in an influential 1997 *New Yorker* article about coolhunters, Malcolm

Gladwell had said something more radical. He said a law of cool was that only certain people knew it when they saw it—namely, cool people, who at the time were making a lot of money spotting the next cool thing for companies. To make matters even worse, he said, "not only can the uncool not see cool but cool cannot even be adequately described to them."[3] If you're a coolhunter, that sounds like a great way to build in job security. But if it's true, it's bad news for any hope of ever figuring out cool. Is cool really like the comic-book character Quicksilver? Does it move too fast to pin down? One thing was clear, though. Cool mattered. It mattered to people, and it mattered to business.

How could something so important be so poorly understood? We weren't going to bet on Gladwell's so-called law that there wasn't a good answer. Our decision to take up Lisa's challenge to look inside the brain for cool was made easier when, a short time later, Alan Alda, who at the time was the host of the PBS TV show *Scientific American Frontiers*, asked whether our lab was working on how hidden motives influence our behavior. A brain scanning experiment could shed light on both these issues, as we were confident that looking inside the brain for cool would uncover unconscious forces that shape cool and its role in our consumption. The folks in Steve's lab were already working on how the brain calculates economic value. But we had been thinking about simple kinds of economic value, such as the value of a snack when you're hungry, and other so-called basic needs. These seemed rooted mostly in the Survival and Habit pleasure machines. Cool was obviously connected to economic value. But what so intrigued us was that it seemed like a different kind of economic value. It didn't seem to satisfy a basic need.

In fact, lots of cool products don't even have any superior functionality. After all, a cool watch doesn't keep time better than an uncool one. Air Jordan didn't revolutionize the athletic shoe industry, or spark riots, stampedes, and violence because of any new coveted technology. Most of the people who bought them didn't even play basketball. Air Jordans were cool, but they revolutionized street fashion, not sports. Ironically, Adam Smith, the eighteenth-century economist most associated with asocial *Homo economicus*, would have understood the allure of Air Jordans. He suggested that our economic life wasn't about just the bare necessities. If it were, we'd stop working as soon as we had food and shelter—just like every other animal. Smith said that what people really worked to obtain was recognition from others—to be observed favorably. Was cool a special sort of recognition of the sort Smith thought drove economies? If it was, it meant that something revolutionary must have happened to the human brain. Your brain keeps track of your social image—your perception of how other people evaluate you. It's doing it all the time, usually outside your awareness. If this part of the brain connected to the brain's reward system, we would experience pleasure when our brain judged that others viewed us favorably. Maybe the evolution of this new connection involved rewiring the Goal pleasure machine (the system that evaluates novel goods). If it did, it would mean that when you see a product your brain computes how much it will likely enhance or hurt your social image. Since economic value stems from pleasure, or utility, this would create a strange new kind of economic value: the *social* value of products. Do we buy some products mainly for their social value—for their cool? Could it be that our economy runs on the currency of cool?

These possibilities raise two kinds of riddles. First, *how* does the brain size up the social value of products? Second, *why* does the brain do this? Answering the *how* riddle requires looking into the nuts and bolts of how the brain responds to cool and uncool products, which we'll explore in this chapter. Answering the *how* riddle will give us clues to the *why* riddle. That is, once we've identified the brain structures involved, we can put them into the larger context of their evolution and then explore what forces shaped them. This will let us see what evolutionary problems were solved by rewiring the Goal pleasure machine for cool. From there, we can start to build a picture of the role this strange sort of economic value plays in our lives.

Having decided to accept Lisa's challenge, we thought about how we could translate her query into an experiment. We knew we needed to capture that "Wow!" feeling you get when you encounter something cool. It might happen when you're channel surfing and an ad for a cool new tablet catches your eye. Or it might be the first time a Tesla car whispers by. You do a double take because some quality about it captures your attention. That's cool. We also suspected that there was something important to capture about the feeling when you encounter the opposite of cool—like the feeling you get looking through awkward family photos or having to go to prom in your mom's minivan.

We first had to figure out how we were going to decide what was cool and what was uncool to show our participants. This is tricky because a teenager and a fifty-year-old aren't going to agree much on what's cool. In fact, what's uncool for teenagers is pretty much whatever's cool for fifty-year-olds. Facebook is one example.[4] As more middle-aged parents use Facebook, their teenage children find it less cool and turn to other forms of social media, such as Tumblr (or whatever replaces whatever replaced Tumblr by the time this book is published). This is no idle migration, as Facebook's value could be challenged by the loss of its cool appeal to teens. That's why Facebook continues to try to refocus its image as a useful utility rather than something cool. Executives at Sony had similar worries about the company losing its cool when it failed to respond to the newly introduced iPod after pioneering the personal music system with its Walkman. Today, Sony's biggest earner is its insurance division, in fact—its electronics division loses money.[5] Losing your cool is risky business, and it's hard to keep because it depends on fickle tastes.

If what's considered cool can change as quickly as the Sony and Facebook cases suggest, doesn't this mean that cool really is like Quicksilver? And doesn't the difference between what a teenager and a fifty-year-old consider cool mean that what's cool just depends on who you ask? These are the sorts of intuitions that underlie the idea that cool can't be defined. There's no question that what people find cool depends on who you ask. It also changes over time. That's why people look back on old photos and mostly cringe that they ever thought neon gym shorts and leggings, '80s hairstyles, and other passé trends were cool. But that doesn't make cool impossible to pin down. To see why, consider the case of beauty.

In the early twentieth century, the actress and singer Lillian Russell was considered the most beautiful woman in America.

Russell weighed two hundred pounds, and her body mass index (BMI) would classify her as either overweight or obese today. In contrast, *People* magazine's most beautiful woman of 2014 was Lupita Nyong'o, who weighs a little over half of what Russell did. But if we could travel back one hundred years in time and scan someone's brain as they looked at Russell and compared it to someone today looking at Nyong'o, we'd see similar brain activity. This shows that beauty isn't just the surface appearance. It's how brains interpret that appearance. At its core, physical beauty is a sign of health: proportions, symmetry, pigmentation, lustrous hair, and so on all indicate health and fertility. In Russell's day, more body fat was a sign of health because food was less abundant than today, and a high BMI was relatively rare. Likewise, wealthy men of her day were known as "fat cats" because they could afford to put on more weight. Today, the opposite is true. In fact, the amount of body fat that's considered most attractive depends on the economic prosperity of countries. In other words, the richer the country, the lower the body fat that's considered beautiful.

Theories of bodily beauty go beyond the surface to extract the deeper and more enduring qualities of beauty: signs of underlying health and fertility. Doing so reveals that despite different forms, there is a commonality about beauty and the response to it that endures over time and across cultures. When the fashion world embraced heroin chic in the 1990s, for example, it played on the deep links between beauty and health by presenting images that challenged them. Models looked emaciated, with dark circles under their eyes, ratty hair, protruding hipbones and collarbones. These stark images were meant to be a kind of anti-beauty. We can approach cool in the same way as we do beauty by seeing how different eras express the same values, such as rebellion, in different ways. In other words, what things are considered cool may change, but the song remains the same.

Based on these considerations, we decided that we should recruit participants who were around the same age as one another and who lived in the same community. That way, we could develop a set of cool and uncool images with that cohort specifically in mind. So we turned to the legendary Art Center College of Design in Pasadena. Art Center alumni have created many iconic television ads, including spots for Nike, Budweiser, Levi's, Coca-Cola, and Mercedes-Benz. The school's transportation and industrial design program produced many of the elite car designers now located in Detroit, Europe, and Japan. Most important, it's a place where cool counts. We first recruited twenty design students to help us make our set of cool and uncool images. We asked them to rate from very uncool to very cool a few hundred images from the following product categories: bottled water, shoes, perfumes, handbags, watches, cars, chairs, personal electronics, and sunglasses. We also decided to include images of celebrities in our database because they can evoke strong cool/uncool responses. Just think of how differently people are likely to respond to the face of Mick Jagger versus that of Barry Manilow (two faces we included in our study). We included four categories of celebrities: male actors, female actors, male musicians, and female musicians.

We chose five specific products within each category that the design students rated cool and five that they rated uncool. We also took care to select products with similar price points, since we didn't want the cool product to also be more expensive. Where items weren't price matched, we typically selected uncool items that were more expensive than their cool counterparts. To our participants, for example, a Kia Soul was a lot cooler than a frumpy Buick sedan that woos buyers in their late fifties, although the Buick costs thousands of dollars more. This raises an important point. While expensive products can be cool, cool products don't have to be expensive. (That's because cool started out as an alternative status system among groups that didn't have access to traditional wealth-based status, as we'll explore in chapter 7.) Even a safety pin turned into jewelry for a piercing could be cool among punk fans. Cool is so especially important among teenagers and twentysomethings because they typically don't have a lot of economic power. But they do have the power to determine who's cool and who isn't.

Once our group of design students had helped us put together our set of images, we recruited another two dozen Art Center students for the brain-scanning phase of the experiment. We made sure they hadn't already heard about it. We also never mentioned that our study was about how notions of cool underlie the ways in which we value a product—in fact, we never mentioned the word "cool." All they were asked to do was to naturally experience the images that would appear on a screen above their eyes once they were settled in the functional magnetic resonance imaging (fMRI) scanner. During this time, Lisa Ling and Alan Alda also came to the lab to take part in the brain-imaging experiment. After Lisa, Alan, and the design students finished the scanning portion of the experiment, we asked them to rate the products in a traditional pen-and-paper survey so we could use what they thought was cool when it came time to look at their brain responses. Then we examined how the brain responds to cool items. The most striking brain response to cool items was activation in a part of the brain referred to as the medial prefrontal cortex (MPFC), which we show in Figure 1.

Before we explain why this result is so intriguing, you may be wondering what a blob lighting up in some part of the brain signifies. *What* are we looking at? Although the technology behind fMRI is extremely complex, the basic principles are pretty simple. Understanding how it works requires just a quick peek into how your brain gets the energy you need to think.

Your brain contains about 80 billion neurons. They process information with one another through connections called synapses.

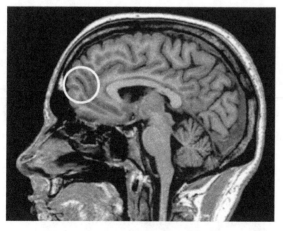

FIGURE 1. The medial prefrontal cortex in the human brain.

Your brain has about 100 trillion synapses, which makes it a kind of computer that's vastly more powerful than even today's fastest supercomputer. And it does this while consuming about 40,000 times less energy than your personal computer. That's pretty impressive. But the brain's energy demands are far from trivial. Your brain accounts for about 2 percent of your weight, but uses about 20 percent of your resting metabolism. In other words, at least three hundred calories a day go to powering your brain, and most of that energy is used to communicate across synapses (otherwise known as thinking).[6] Since finding that much energy before the advent of modern conveniences wasn't a simple matter, feeding such an energy-hungry organ was a major evolutionary hurdle. One evolutionary response to these challenges was to tightly couple thinking and blood flow, since blood flow delivers the oxygen and glucose that powers your brain. So thinking increases blood flow—not to the entire brain, but just to the small patches of synapses underlying the particular thought.

An fMRI machine is essentially a detector of these very small changes in blood flow. It works because oxygenated and deoxygenated blood differ in their magnetic properties, and the machine can detect these fluctuations. So what we're looking at in a typical fMRI image is the changing pattern of blood flow in response to some task. Blood flow isn't thinking. That's what's going on at synapses. But neuroscientists can't measure synaptic activity *directly* in humans because that would be invasive—electrodes would have to be inserted into the brain (this is only done in some rare cases involving epileptic patients prior to surgery). A big upside of fMRI is the fact that it's completely noninvasive. Using blood flow as an indirect signal of synaptic activity is the best we can do today. It's proven to be a flexible experimental framework—participants in fMRI experiments can play games, watch and listen to videos, taste wine, make decisions, or even have an orgasm.[7]

Before fMRI emerged as a workaday technology around 2000, we knew almost nothing about the MPFC (the medial prefrontal cortex we mentioned earlier). Back then, the area of the MPFC we'll focus on was referred to as Brodmann area 10. The German neuroanatomist Korbinian Brodmann divided the human brain into fifty-two areas in the early twentieth century. He did so based on how different areas of the brain looked under a microscope. It was extraordinary work, but it wasn't at all clear whether these areas meant much in terms of brain function. By the late 1990s, when the "Decade of the Brain" was about to come to an end, many scientists argued that we had learned more about the brain in that decade than in all previous ones combined. But we still didn't know much more about area 10 than what Brodmann knew almost a century earlier. (Neuroscientists liked to joke that area 10 was the brain's Area 51.)

In fact, arguments about what distinguished the human brain from that of the chimpanzee still relied heavily on the work of classical neuroanatomists of the early twentieth century. After accounting for differences in body size, had every region of the human brain expanded compared to a chimpanzee's brain? Had only some parts expanded while others shrank? Or were we just another ape in terms of brain size?

According to one popular view, the entire front third of the human brain, the prefrontal cortex, had expanded. This fit well with the image of *Homo sapiens*, the wise man, whose evolutionary distance from other apes was counted off in terms of reason and rationality, as we thought that the prefrontal cortex was the seat of higher reason. It wasn't until 1998 that Katerina Semendeferi and her colleagues used MRI to collect structural scans of different primate brains to make new comparisons. They discovered that the relative sizes of human and chimpanzee frontal lobes (in proportion to total brain volume) were nearly identical.[8] When they compared Brodmann area 10 across seven primate species, they discovered something quite intriguing.[9] It was twice as large in humans, in relative terms, than in other primates. It appeared to be the only region of the prefrontal cortex that was expanded. The only problem was, no one knew what it did.

In retrospect, it's not too surprising that the region of the brain that appeared to underlie human uniqueness remained one of the last to be explored. Up to this time neuroscience had relied largely on animal studies, and it's obviously difficult to study uniquely human capacities in animals. For this reason, the development of fMRI was especially important. Indeed, as neuroscientists were developing fMRI and related technologies, they accidentally discovered tantalizing early clues into the functions of the MPFC (brain imaging no longer uses Brodmann areas, so we'll follow this convention and refer to this area as the MPFC from now on). As we know, fMRI detects changes in blood flow, so experiments depend on comparing blood flow when the brain is doing different things. The most obvious way to do this seemed to be to compare blood flow when someone was thinking to blood flow when the brain was "relaxed." Imagine being asked to look at a complex image and then to relax with your eyes closed in between the images. Intuitively, it seems like the "relax" condition would be a period of relatively quiet brain activity. Yet when researchers examined these patterns in a variety of experiments, they discovered that the brain at rest was typically more active than it was during the demanding task. What was going on?

If you close your eyes and relax for a few minutes, chances are you'll start daydreaming. You'll start thinking about what you have to do in the next few hours and maybe even start replaying earlier events. In other words, unless you're an experienced meditator who can clear her mind of mental clutter, when you disengage your brain from the world, you spend that time in an internal narrative of planning and ruminating. In one instance, you might recall an event from your morning—perhaps a friend said something that stuck with you. You mull it over in your mind, replaying the episode over and over again. Maybe you work out a plan for how you'll respond to him. During all this time, you reflect on yourself and imagine yourself interacting with others by replaying earlier encounters.

These early fMRI experiments suggested that the MPFC was involved in daydreaming, planning, and ruminating. One thing all these have in common is that they involve reflecting on yourself. When you're ruminating, you're recalling something from your personal past, like that episode with your friend. These are your memories and they help define your sense of self. Neuroscientists refer to these sorts of memories as episodic memory. These are different from your memories about impersonal facts, such as your ability to recall the capital of a state. Some neurological patients have lost all their episodic memories but their semantic memories are perfectly intact.[10] They can tell you the name of their high school, but they can't recall their graduation. They can remember the date of their anniversary but can't remember their wedding. When you're planning, on the other hand, you're thinking about your future self. Maybe you're thinking about going on a vacation and picturing yourself relaxing on a beach somewhere. Patients who have lost their episodic memory also can't plan for the future. They can't imagine their future self. These terrifying cases reveal the intimate connection between the MPFC and our sense of self.

Did the fact that cool products activated the MPFC in Lisa, Alan, and the design students suggest that cool was intimately related to their sense of self? In what ways could cool relate to our self? To answer that, we need to look deeper into the functioning of the MPFC. Soon after the finding that the MPFC was involved in daydreaming, planning, and ruminating, researchers began exploring the MPFC's role in our sense of self in more depth. One way to do this was by asking people whether certain adjectives described them. Consider, for example, what happens when we ask you whether you're honest. To answer, you need to look inside yourself. Over and over again, tasks like this one consistently activated the MPFC in study participants.[11] These results were intriguing, but our experiment on cool didn't involve asking people to explicitly look inside themselves. If Alan, Lisa, and the design students' perception of a product as cool involved relating it to their sense of self, it would mean that our sense of self can be activated even without our conscious, or explicit, awareness.

This might sound paradoxical at first, but there's a difference between self-reflection and self-relevance. For example, you've probably been at a packed cocktail party when suddenly you hear your name in a conversation going on across the room. In the sea of conversations going on, your name wasn't any louder than any other words. In fact, it might have been said in a hushed tone if the people didn't want you to hear. But you heard it because it's self-relevant. Chances are you weren't vigilantly scanning the room in the off chance someone was talking about you. Instead, your unconscious brain was monitoring the room for you. In short, although you're typically not aware of it, your brain is keeping track of how relevant things in your environment are to you—from your name to significant products. Think of your unconscious self-relevance detector as a dial that rises and falls as you navigate your world. When the dial reaches a self-relevance threshold, it shifts your attention to whatever triggered it, like overhearing someone talking about you.

Whether or not something is self-relevant depends on whether it reflects our sense of self, the internal model we have of ourselves, including how we perceive our personality, tastes and preferences, physical characteristics, skills, and so on. If you've ever taken up a new hobby like cycling and then started seeing cyclists everywhere, it's not because you started a trend that everyone's copying. It's because cycling is now self-relevant to you. Likewise, products that reflect your sense of self will be more self-relevant to you. That's why only some products grab your attention when you're shopping or leafing through a magazine. In our experiment, this is why Alan, Lisa, and the design students' MPFC activation was greater for products they gave the highest cool ratings to after the brain scan. Their MPFC was tracking how relevant products were to their sense of self.

Self-relevance seems to be an important part of cool, but it's not the whole story. Were Alan, Lisa, and the design students' brains responding to something else as well? A clue that they were stems from research revealing that the MPFC is involved not only in thinking about ourselves, but in thinking about other people. Most of us think about other people so naturally that we don't realize how complex a process it is. In fact, we may be the only species capable of doing it. To see what we're getting at, consider why the old joke about a chicken crossing the road is funny. Asking why the chicken crossed the road invokes a special capacity in us called theory of mind. This is our ability to think about other people—and sometimes other animals and objects—as having minds. We naturally explain their behavior as the result of their beliefs, desires, hopes, fears, and so on. We expect the punch line to the chicken joke to mention something about the chicken's mind: perhaps it was afraid of a farmer, or it longed to see the world. The humor comes from the failed expectation that our theory of mind creates. As natural as this ability is for most of us, some developmental disorders, such as autism, involve disruptions in the development of theory of mind. For some people with autism, other people seem more like physical objects than persons with minds. As a result, they struggle with social interactions.

These two functions of the MPFC—thinking about ourselves and thinking about others—may seem to be distinct capacities, but they are related in ways that are critical for understanding the social dimensions of consumption. Our self-concept isn't solipsistic. The hypothetical *Homo economicus* may never be influenced by other people, but real humans develop a sense of self over the first two decades of life through interacting with others, and using those interactions to construct our self-concept. And because we can think about other people, we can think about what they think about us. Over time, we internalize this social feedback into our own self-concept.[12] The role of the MPFC in thinking about what other people think about us will turn out to be the crucial piece of the puzzle for understanding the pattern of brain activation we observed in Alan, Lisa, and the design students.

Indeed, discovering the functions of the MPFC helped overturn the notion that the evolution of the human brain was primarily about the growth of abstract problem solving. What was so crucial for human evolution was the emergence of our ability to think about ourselves and others in terms of minds and mental states. This allows us to create complex social arrangements because we use our theory of mind to navigate our social worlds. Just think how often during the day you wonder what someone else is thinking, whether they are angry with you, whether a comment you made embarrassed them, even whether a certain look a coworker gave you was a flash of contempt or an attempt to be funny. All these apparently incidental details of everyday life couldn't take place without theory of mind. And just think of how complicated our theory of mind becomes when we start bargaining and negotiating with others. Our social world would come crashing down if we lost these abilities. As Adam Smith suspected, our consumption isn't about the bare necessities. It's part of our social world, made possible by the MPFC and theory of mind.

From Self to Sociometer

The fact that our self-concept draws on how we think others think about us presents a tremendously intriguing possibility. It suggests that in our experiment with Alan, Lisa, and the design students, their response to a cool product involved a calculation—likely unconscious—of how others might think of them *with* that product: that is, how the product might enhance their social image. This calculation is critical to human consumption, as Adam Smith argued. But thinking about how other people think of us plays a pervasive role in our social lives beyond consumption. Indeed, it is a key to creating our complex social structures, and the logic of its evolution will provide essential insights into why we consume. To consider this, imagine you're at a restaurant across town. You excuse yourself from your guest at the table to visit the restroom. You enter the restroom as a stranger is on his way out. As he looks up and sees you, you notice that his path to the door subtly changes course and he heads toward the sinks. You exchange brief smiles, but you detect the telltale signs of embarrassment on his face. As inconsequential as your encounter with this stranger may seem, unraveling what took place uncovers some of the deepest mysteries of being human and the forces that drive our consumption and animate our economy.

To begin, let's start with some facts. It's a fact that people in public restrooms are more likely to wash their hands when in the presence of others, as this stranger's subtle detour to the sinks reveals. Moreover, in telephone surveys people lie about their restroom hand-washing habits: 96 percent claim they do, but a discreet observational study of more than 6,000 adults in public restrooms reveals that only 85 percent actually do so.[13] Why are people more likely to wash their hands when others are around—and why would some people lie about their hand-washing habits to a stranger on the phone? The answer lies in a radical transformation to the brain's value systems somewhere in our evolutionary past—a game-changing transformation that makes human economic life possible.

Our starting point for sorting out this mysterious transformation is the fact that the person you encountered in that restroom felt embarrassed, and this discomfort caused him to change his behav-

ior to alleviate his embarrassment. We've already discussed the connection between emotions and rewards and the idea that utility is
cashed out as an emotional brain state. In those terms, the stranger
felt disutility as a result of embarrassment, or the anticipation of it,
which motivated him to act. The link between emotions and reward
is what makes us pleasure machines. But embarrassment is a strange
emotion. The emotions we've looked at so far relate to basic sorts of
rewards. The positive emotion we receive from eating food, for example, seems pretty straightforward. At its core, it's probably not
too different from the pleasure other animals receive from food as
well. It's also not too complicated to see how we could explain the
logic behind why food is linked to pleasure in terms of evolution.
But embarrassment doesn't seem to relate to basic sorts of rewards
at all. Some emotions are only triggered when we think, perhaps
subconsciously, about how others are thinking about us—which was
made possible by the evolution of the MPFC. These are the social
emotions, emotions such as embarrassment, guilt, shame, and pride.
Think about it for a moment. How could you feel pride if you didn't
sense that others valued you or held you in esteem? Similarly, the
person in the restroom could only feel embarrassment if he sensed
your evaluation of him. When the stranger saw you in the restroom,
his brain registered the risk of negative social evaluation by another
person witnessing him transgressing a social norm. Little of it was
likely conscious—it happened nearly mid-step—but it resulted in
an aversive emotion that altered his behavior. The social emotions,
then, involve perceptions of other people's evaluation of yourself or
a facet of yourself. It may sound straightforward, but we're likely the
only creatures to have this capacity.

Most of us appear to be exquisitely sensitive to the approval and
disapproval of others. Consider that 75 percent of the population
suffers from glossophobia, the fear of public speaking. Many people
even rank this fear as high in severity as the fear of death. Public
speaking is, of course, an opportunity to embarrass yourself in front
of a crowd—to discover that your favorite joke is in fact not very
funny, that you really shouldn't have relied on your memory, or, the
next day, that your flop of a speech has become the butt of the latest
water-cooler conversation. Poor Miss South Carolina 2007 became

a media sensation when she flubbed her answer to why so many Americans can't locate the United States on a map—nearly 100 million people have watched her performance on YouTube. Howard Dean's infamous 2004 Iowa speech instantly became fodder for comedians, who referred to it as his "I have a scream" speech, and it likely contributed to the end of his presidential candidacy. Indeed, YouTube has become a graveyard for bad public speeches, including Phil Davison's speech making his bid for nomination for the office of Stark County, Ohio, treasurer in September 2010, which made him a viral sensation in part because of the vicarious embarrassment we feel watching others flub so monumentally. Even some Air Force pilots, trained to fly perilous missions, choose not to become commercial pilots due to their overwhelming fear of having to make in-flight announcements. When asked to recall a traumatic event in their lives, most people recall episodes involving embarrassment, from public flatulence and vomiting to slips and falls. There is no doubt that embarrassment can leave emotional scars.

Again, the MPFC appears to play a critical role in the link between the self and social emotions, which we can think of as responses to perceived evaluations of the self by others. It's important that we keep in mind that these evaluations are subjective—that is, they are perceptions of how others evaluate us. We suspect that differences in sensitivity to social evaluation play a central role in social behavior and consumption, a theme we'll return to in later chapters. The link between the self and social emotions in the MPFC suggests that the MPFC may track moment-to-moment changes in perceived evaluations of oneself. Although Francis Edgeworth fancied the hedonimeter as a device capable of measuring physiological pleasure, we can also think of the brain's value signals as a kind of internal hedonimeter that measures its internal currency. We'll therefore use the term "hedonimeter" to refer to this internal monitoring of value representation.

Here's how we want to extend this idea: since basic emotions encode the values of a hedonimeter, social emotions may encode the values of an internal sociometer, which tracks perceived social valuation.[14] Think of this as an internal monitor that measures your perceived social approval, a meter whose "readings" we experience as various social emotions, such as pride, shame, embarrassment, and guilt. Just as your hedonimeter is constantly evaluating your environment, your sociometer is evaluating your social environment, often beyond your awareness. Remember that this evaluation is subjective—it is your perception of how others view you. This is important because economic value is likewise subjective, and we believe the sociometer tracks a fundamental kind of economic value: social currency.

An intriguing property of the MPFC for how our sociometer takes shape is the particularly protracted development of this part of the brain, which continues through adolescence. As parents of teenagers can tell you, with the onset of puberty their kids become increasingly preoccupied with peer evaluations, a preoccupation that grows in tandem with an intensification of social interaction and an expanding self-concept. In one study, children, adolescents, and adults took part in an fMRI experiment in which they were told they would be viewed by a same-sex peer through a video camera at various times during the scanning.[15] A light would tell the participants when the camera was on. Even the thought that a peer was watching them was enough to induce self-conscious emotions, which were mildly aversive and which rose in magnitude from childhood to adolescence, peaking at around fifteen years of age and then partially subsiding in adulthood. Corresponding to these differences was a large rise in the engagement of the MPFC among adolescents compared to children, suggesting that the MPFC is involved in the heightened emotional value and increased self-relevance that social evaluation has for adolescents. (Interestingly enough, it's adolescents who become extremely sensitive about how peers perceive them, and adolescence signals the onset of caring about cool as well.)

If being under the watchful gaze of others is enough to spark self-conscious emotions, imagine getting feedback from a stranger who watches a video of you being interviewed. This was the scenario of a brain-imaging experiment by Naomi Eisenberger and her colleagues, along with Mark Leary, one of the proponents of the sociometer theory.[16] After participants were interviewed by the scientists, their brains were scanned while they watched a screen displaying words such as "annoying," "kind," "serious," "boring," and "sincere," representing the impressions of another participant who was watching the taped interview. They found that participants had very strong responses to negative feedback, which resulted in greater activation of the MPFC than did positive feedback. The experiment also activated the anterior insula, a region of the brain involved in processing both physical and social pain. This result emphasizes that the MPFC is not itself a positive or negative reward center, but one involved with measuring self-relevance. Its connections to centers of positive reward (striatum) and negative reward (insula) form central elements of the brain's networks for positive and negative social emotions. Since we find negative feedback more salient than positive feedback, it's reasonable that negative feedback would result in greater activation of the MPFC and would likely initiate greater mentalizing, as it has been shown in a variety of studies that negative feedback is ruminated on more thoroughly than positive feedback.[17]

So far, we've been discussing the idea of a sociometer and social emotions with an emphasis on negative social emotions. Let's take a look at positive social emotions and the sociometer by considering whether the following is an accurate statement about you: "I wouldn't hesitate to go out of my way to help someone in trouble."

Think about it for a moment, and then answer to yourself. Now suppose we ask you the same question, but with a twist: This time, you'll reveal your answer to two strangers who are watching you in the next room via video. If we were monitoring your brain activity in these two contexts, do you suppose we would see any differences, even assuming your answer didn't change? It turns out that when you reveal answers that show you in a positive light in the presence of others, your MPFC activates more strongly than it does when you answer to yourself.[18] The striatum, the reward-processing structure, is likewise more activated in the presence of others. These links between MPFC and striatum indicate a strong connection between perceived social evaluation and reward. In fact, simply finding out that someone you recently met likes you activates these same brain regions, including the MPFC and the striatum. Still more intriguing, the strength of this activation is even higher if the people who like you are those whom *you* regard highly as well.[19]

An Economy of Self-Esteem?

The links between MPFC, striatum, and social evaluation are astonishing. Perceiving that others hold you in esteem affects your sociometer in much the same way that more basic rewards impact the hedonimeter. The flurry of activity in the bee's brain as it homes in on a flower it predicts holds nectar; being handed a $20 bill—both impact the hedonimeter. The bee needs nectar to survive. You can buy yourself goods with the $20. But you can't eat social approval. It satisfies no "basic" needs. And yet our brains treat social approval like basic rewards. Indeed, if we gave you $20 and offered you the chance to pocket it or to donate it to a charity, we could likely influence your choice by giving you the opportunity to make your donation in front of other people. You might think that losing $20 would result in a silencing of your brain's reward circuitry, or even activate punishment circuits registering the loss. Instead, your reward circuitry would activate strongly in this scenario, indicating that your brain calculates the social approval the scenario affords as more valuable than the potential monetary gain.[20]

For the human brain, social approval is a kind of currency, a type of economic value that substantially overlaps with the same basic reward structures involved in monetary rewards.[21] But the MPFC and its sociometer are critically required to register perceived social approval. This abstract good—social approval, reputation, esteem, or status—plays a central role in our motivation and behavior, and it is the currency that drives much of our economy and our consumption. As seen in the subjects who gave up $20 to enhance their reputation, a central dynamic of our consumption lies in converting money into perceived esteem. We will see how this brain-based quest for esteem gives rise to our uniquely human form of consumption, including the permeation of our economy by social values such as cool, and how it underlies a fundamental facet of being human.

Now, let's examine the link between the sociometer and reward in more detail. A variety of terms have been used to describe what the sociometer measures, but the best-known one is "self-esteem." Keep in mind that although "self-esteem" sounds like a private self-evaluation, sociometer theory suggests a tight link between self-esteem and perceived evaluation by others. Positive self-esteem has long been regarded as a fundamental human need. Indeed, although the aphorism "Know thyself" has long aligned wisdom with accurate self-knowledge, a number of psychological biases appear to make a positive self-concept more of a priority than an accurate one.[22] For example, we tend to think of ourselves as better than average, we tend to prefer the company of people who reinforce our positive self-image, we attribute our successes to our own personal traits but our failures to others or the situation, and we even think of ourselves as less biased than others. (Depression is often associated with a reversal of these processes, including overattributing bad events to the self, focusing on negative information that matches a negative conception of self, and dismissing positive events as not being due to the self.)

For these reasons, opportunities to boost self-esteem are highly motivating and rewarding to most people. In a surprising series of findings, researchers discovered that college students value boosts to their self-esteem, such as being praised or receiving a good grade, more than eating a favorite food, receiving a paycheck, seeing a best friend, or engaging in a favorite sexual activity.[23] Indeed, the feeling of pride is a universal social emotion that involves positive feelings of self-worth. Even prideful body language is universal. When you feel pride, you tend to expand your chest and posture and tilt your head up (literally expanding yourself). Athletes across many cultures raise their hands above their heads when they perform well—again, a physical expansion. This victory salute is nothing new—there are many ancient Greek statues of victorious athletes with their arms raised high. In contrast, when people feel shame, they narrow their chest and slump their shoulders, shrinking the body. These nonverbal displays of shame are likewise universal across cultures in both adults and children, and are related to the submission displays of nonhuman animals as well.

What does receiving a compliment have in common with the products Alan, Lisa, and the design students found cool? If we just looked at their pattern of brain response without knowing what caused it, we might have thought it was caused by positive social feedback such as a friendly gesture, which generated a positive social emotion, like pride. The fact that a product they saw as cool caused it reveals that products have a social life. In other words, the level of cool of a product registered on their sociometer. Lisa, Alan, and the design students valued the cool products for the imagined effect on their self-esteem, their social image: what they thought other people would think about them if they were seen with that product.

The ability of a product to enhance self-esteem is an economic game changer. Like other forms of economic value, the social value of products is subjective and depends on an often implicit and complex evaluation. But just as our hedonimeter is the source of economic value for things like food, our sociometer creates economic value when it responds to cool products. Human desires create economic value. For Lisa, Alan, and the design students, their desire to be observed favorably by others imbued the cool products with economic value. The social emotions of the sociometer—no less than the more basic emotions of the hedonimeter—create economic value, making our economic life irreducibly social.

The Red Flag of Stodgy Loafers

So far, we've been considering positive responses to cool, but how did Alan, Lisa, and the design students respond to products they found uncool, like NASCAR sunglasses, a frumpy chair, a Buick, or a Bunn-O-Matic home coffee machine? We found that the MPFC also activated in our participants when they responded to uncool items. As we know, negative social feedback reliably activates the MPFC. In fact, negative information about oneself is typically more salient than positive information.[24] Negative social evaluations produce strong aversive emotions, such as embarrassment and shame, which track the loss of self-esteem. Just the thought that someone is watching us is enough to produce uncomfortable self-conscious emotions. The finding that our participants responded to uncool products in terms of negative social emotions suggests another interesting connection between products and social norms.

A series of findings have revealed that reactions to the violation of social norms likewise crucially involve the MPFC. Imagine, for example, that you are at a friend's house for dinner. You take a bite of food, choke on it, and cough it out. This would be an embarrassing social situation, but it would be an unintentional violation of a social norm, since you didn't mean to spit out your friend's food. In contrast, suppose you take a bite, but don't like your friend's food, and so spit it back onto your plate. In this case, you've intentionally violated a social norm. Sylvie Berthoz and her colleagues examined these scenarios, and found that the MPFC was activated when the stories portrayed people (either "you"—the study participant—or a third party) both unintentionally and intentionally breaking social norms, but that the activation was more pronounced for intentional social norm violations.[25] This result suggests that intentional social norm violations are also seen as more self-relevant. Even when they involve a third party, they make us think more about ourselves and how others would react if we committed such violations, for which we would be held accountable.

Uncool products also activated the insula in our participants. The front part of the insula, referred to as the anterior insula, is a brain region of intense interest to scientists, as it appears to be involved in a wide variety of human experiences, including craving, pain, empathy, disgust, social rejection, humor, sexual stimulation, trust, romantic love, anger, fear, sadness, and risk. Work in Steve's lab, for example, found that activity in the anterior insula increases as the risk of a gamble increases—the riskier the gamble, the greater the insula activation.[26] Although we've enumerated what looks like a laundry list of emotions, some positive and some negative, corresponding to insula activation, all of these emotions involve a strong bodily sensation. Notably, when we talk with people about uncool products, we're struck by the language they use to describe their reaction to those products. Words like "shudder" are common, suggesting that the thought of being associated with an uncool product can cause people to tremble with fear or revulsion. So insula activation in response to uncool items likely underlies the strong negative emotions we feel when the sociometer calculates a potential loss to our social esteem—a fear that is at the heart of social phobias, from fear of embarrassment to the terror of public humiliation.

Indeed, there are fascinating links between the anterior insula, emotions, and the violation of social norms. Consider, for example, the Ultimatum Game, an economic task that has been used by economists, psychologists, and anthropologists throughout the world to probe how fairness norms work in various societies.[27] The rules of the game are straightforward. In a typical version of the game, a person is given some money and is told that they must offer some portion of it to another person, who will decide whether they want to accept or reject the offer (the second player knows how much money the first player was given). Both players are told that if the player accepts the offer, then both players keep their portion. However, if the player rejects the offer, then both players get nothing. Typically, the game is played anonymously so the players don't know who they are playing with.

The person making the offer must therefore think about what the other player will consider a fair offer. If the hypothetical economic decision maker *Homo economicus* played the ultimatum game, he would offer the smallest unit of currency available to him. He would assume that the person receiving the offer was rational, and a rational person would accept even a penny over nothing. In reality, most people angrily reject such lowball offers. In fact, activity in the anterior insula in response to being treated unfairly in the ultimatum game (when someone offers you $1 out of $20) predicts the likelihood that you'll reject that offer.[28]

Inside Alan's and Lisa's Brains

As our experiment revealed intriguing patterns of brain activation in response to both cool and uncool products, we next turned to the question of individual differences. On the one hand, our results suggest that cool products act like a "carrot" by attracting people with the lure of boosts to self-esteem. On the other hand, uncool products act like a "stick" by repelling people with their threat of diminished self-esteem. Since we know that some people are motivated more by carrots than by sticks, and vice versa, we wondered whether people differ in whether they are more sensitive to the positives of cool or the negatives of uncool. If so, it suggested that similar consumer decisions might stem from very different motivations. For example, two people may buy the same cool new tablet or smartphone, but one may have been motivated by the lure of its cool, while the other may have been motivated more by the desire to avoid an uncool alternative. Do these differences play a role in consumer decision making?

As we examined how our participants responded to cool vs. uncool products, we found intriguing differences. In fact, Lisa Ling and Alan Alda represented two very different sorts of responders. Lisa was more responsive to the cool products than the uncool ones. When we met up with her to discuss her results as part of her television episode, we started out by asking her about her shopping habits and whether she would describe herself as a shopper. She admitted that she would, and told us about her handbag and shoe expeditions. She even asked the names of some of the products she saw in the scanner, as she'd developed a liking for them and wanted to hunt them down. We then asked her whether she'd describe herself as impulsive. She laughed and asked us whether we'd consider buying a car sight unseen over the phone as impulsive. Alan, on the other hand, was more responsive to the uncool products. Like some other participants who showed more sensitivity to the uncool products, he also rated more of the products as uncool than did people with responses like Lisa's. Did Alan's brain find the uncool more salient than the cool? What other responses to cool and uncool might there be among people, and how might these differences play a role in consumer decision making and behavior?

To examine these questions, we turned to research on individual differences. In particular, we wanted to know: How do people differ in their sensitivity to reward and punishment? How do people differ in their sensitivity to social approval and disapproval? One way to approach these questions is by picturing a Dr. Dolittle–like pushmi-pullyu in your brain, the Behavioral Activation System (BAS) and the Behavioral Inhibition System (BIS).[29] One system pushes you toward reward and the other pulls you away from punishment. Rewards arouse your BAS, which ramps up your motivation to get them. You've no doubt felt the emotions your BAS inspires, including elation, happiness, and even hope. In contrast, your BIS steers you clear of punishment. Your BIS inspires anxiety, fear, and sadness. Personality differences might stem from differences in the strengths of these two systems in people. A person with a stronger BAS is not only more sensitive to reward but also may be more extraverted and impulsive. Someone with a stronger BIS is not only more sensitive to punishment but also may be more anxious and neurotic. Furthermore, attention-deficit/hyperactivity disorder (ADHD) may be due to a high BAS, while depression may be related to a very low BAS. The higher a subject's BAS score, the greater their ventral striatum activation to a monetary reward. Conversely, the higher a subject's BIS score, the lower their ventral striatum activation to a monetary reward.[30] This is a classic example of differences in reward sensitivity: the same monetary reward has different hedonic effects depending on differences in these systems. In an influential study, Turhan Canli and his colleagues found that the personality dimensions of extraversion and neuroticism related to differences in brain activation to either positive or negative emotional scenes, further illustrating that individual differences in brain function color how individuals see the world.[31]

Based on these considerations, we developed a framework for individual differences in our experiment and how they may relate to social consumption.

Using the four major categories in the BAS/BIS system, we outlined four major consumer types (see Figure 2 on page 86). High positives (high BAS), who tend to be extraverted, impulsive, and highly sensitive to social reward and influence, would show heightened activation in MPFC in response to cool items. Interestingly, compulsive buyers tend to score high on BAS scales as well.[32] Low positives (low BAS) are less sensitive to reward and presumably less sensitive to social influence. High negatives (high BIS) tend toward high anxiety and are highly sensitive to social punishment and social exclusion. Such individuals would show heightened activation in response to uncool items. Interestingly, people with generalized social phobia show extremely high MPFC activation to social norm violations, particularly unintentional ones that trigger embarrassment.[33] Low negatives (low BIS) would show reduced sensitivity to social punishment.

It's worth bearing in mind that although we've conceptualized consumers in terms of four major categories, the reality is far more complex in two main ways. Although we've used discrete categories,

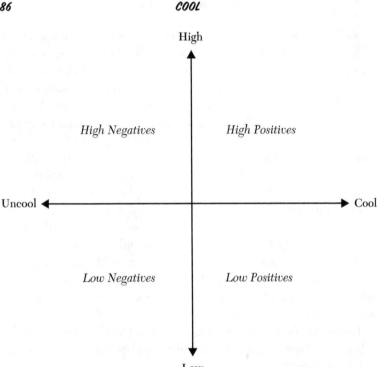

FIGURE 2. Four categories of consumers based on differences in responses to positive (cool) or negative (uncool) products and amount of activation to those products.

individual differences represent a spectrum. In addition, there are likely to be more complicated combinations, such as the potential for an individual to be more sensitive to both rewards and punishments. This framework is useful for organizing individual differences, and our main use of it is heuristically to consider how individual differences in reinforcement sensitivity give rise to different consumer behaviors. For example, we have an ongoing interest in whether high positives tend to be early adopters. Do high negatives tend to be more conformist and later adopters, since they would tend to take less risk in their consumer choices? We also think that low negatives are less sensitive to social influences, and so may not invest as much energy in social consumption. They may be more swayed by the functional features of products, and may, for example, choose a minivan for their family despite its uncool stigma, which has driven a long, steady decline in minivan sales.

Our experiment involving Lisa, Alan, and a group of design students had uncovered a surprising link between our brain's adaptations for social life and our consumption. Cool turns out to be a strange kind of economic value that our brains see in products that enhance our social image. But our experiments hadn't yet singled out exactly in what ways our social image is enhanced by association with cool products. It wasn't just the status associated with the cost of products, since our participants thought that many less-expensive items were cooler than more-expensive alternatives. This leaves us with some pressing questions: How does cool enhance our self-image in different ways than luxury does, and why did our brains evolve to link our self-image with feelings of pleasure that create economic value? Is this just misbegotten vanity, or is there a deeper truth lurking in these results about who we are, why we consume, and why cool is intimately linked to economic value?

4

CONSUMER EVOLUTION

You're at the mall on a busy Friday night when a woman wearing a plain green sweater approaches you, clipboard in hand. She makes eye contact with you and asks if you have time to answer a brief survey. How likely are you to agree to answer her survey? Would your decision be swayed by whether her sweater had a Lacoste alligator logo on it? How about a non-luxury logo? Everyone we've posed this question to is sure their behavior wouldn't be influenced by that little alligator or another luxury logo. But the facts tell a different story. Whereas only 14 percent of people agree to answer a survey when approached by a woman wearing a plain sweater or one with a non-luxury logo, a whopping 52 percent agree to answer the survey when an alligator logo is added to the sweater.[1] Now suppose that woman just knocked on your door asking for a donation to a well-known foundation devoted to raising money for research on heart disease. Would an alligator on her sweater influence how much you donated? Everyone we asked said it wouldn't. But the fact is that people's donations almost double. That alligator may be small, but its effects sure aren't.

Explaining the "alligator effect" goes to the core of the question of *why* we consume. As we suggested in the last chapter, explaining *why* we consume requires placing it in a larger evolutionary context to explore what forces shaped the brain structures underlying consumption. This evolutionary investigation will lead us to a radically different view of what underlies our economic behavior and shapes consumer culture than the stock views we've seen so far. It will reveal not only why those traditional views can't explain the alligator effect, but why they seriously distort the nature of our consumption. We'll come to see that modern human consumption emerges out of ancient evolutionary adaptations to the central problems of social life.

Let's start with some theories of consumption to see how they fare when it comes to explaining the alligator effect. Sociologists such as the late Pierre Bourdieu suppose that our need for distinctiveness drives our consumption.[2] The sociologist Niro Sivanathan supposes it's our need for self-affirmation.[3] The marketing professor Aric Rindfleisch links our consumer motivation to our awareness of our own mortality and how that awareness creates a psychopathological love of that death-denying abstraction, money.[4] These assume our consumption is irrational at best and probably pathological. So they contrast these consumer motives with "genuine needs and desires."[5] They blame our purchase of a Lacoste sweater on extrinsic motivations, such as the desire for fame, as opposed to intrinsic ones, such as the desire for community. Even putting aside the problem that the distinction between intrinsic and extrinsic motivations is more ideological than psychological, there's a more pressing problem with these theories. What's important about the alligator effect is the logo's influence on other people, not directly on the wearer.

Status competition theories can't explain the alligator effect either. Recall that these theories assert that consumption motives lie in the desire of consumers to distinguish themselves from others. There's a reason why critics who believe in Veblen's ideas of conspicuous consumption talk of "snob effects." Since snobs get satisfaction from viewing themselves as above others, the motive behind conspicuous consumption is a certain snobbish pleasure. An expensive logo's wearer looks down on you and tries to make you feel inferior. These critics predict that observing that woman wearing a Lacoste sweater—or any other instance of conspicuous consumption—ought to invoke in you a blend of humiliation, envy, and one-upmanship. In the game of status competition, that sweater is the consumption equivalent of someone trying to slam-dunk on you, sprint by you at the finish line, or catch a game-winning touchdown over you. That sweater ought to invoke a resentment of its owner in you and should trigger your consumption motive to alleviate the indignity of perceived inferiority. So the effect of status competition is to increase social distance between consumers—the snob and the object of his or her derision. The problem is, the alligator's effect increases cooperation between its wearers and the people they interact with, making those people both more likely to socially engage with the wearers and more likely to be more prosocial—for example, contributing more to a wearer's charity. The problems with these theories only get worse when we turn to another experiment. For this one, we're going to ask you to play a game known as the Dictator Game. We'll start by giving you twenty one-dollar bills. You have one decision to make and then the game ends. That decision is whether to give some portion of the $20 to another person, whose photo you see on a computer screen. The person lives in a faraway city, doesn't see you, and no names are given, so the two of you won't interact in the future. We can even design the game so the experimenter running it won't know how much you give. The other person has no say in your decision to give or not to give any money—it's entirely up to you. Even when the experiment involves strictly anonymous interactions, people give, on average, $5 to the recipient.[6] This in itself is intriguing, since *Homo economicus* would keep all the money for himself. But suppose the person you see on the video screen is wearing a shirt with an alligator on it. Do you think the presence of that logo would influence how much you give? Well, just as in the example above, people give about 25 percent more money to a recipient wearing a luxury-brand shirt than to one wearing a plain shirt. In fact, it's the exact same picture, except in one the logo has been digitally removed. So that 25 percent increase is due solely to the alligator effect.

Here again the alligator effect is a large one. What's more, it seems initially counterintuitive. Why give more money to someone who can afford luxury goods? Wouldn't they seem to need it less? What about the idea that you should feel animosity toward the recipient in the alligator shirt? The Dictator Game seems like the perfect opportunity to vent your resentment and give them nothing, since they can't retaliate. The fact that people give more—not less—is a powerful clue that there is something fundamentally wrong with the standard accounts of consumption.

The stock theories of consumption we've examined so far are catastrophic predictive failures. As we'll discover, the fatal flaw of status competition views is that they rely on outmoded evolutionary thinking (other theories don't make a connection to evolution at all). Status competition isn't simply a vestigial arms race for bigger and bigger trophies. This view of status competition can't be the right story because it leaves out an essential evolutionary insight: We don't compete simply for the sake of competing. *We compete to cooperate.* That might sound like a subtle difference, but it makes all the difference in understanding why we consume and how an economy of cool consumption emerged. For that, let's build a theory of consumption on the bedrock of an evolutionary force that standard theories overlook—social selection—and see why it's a force that changes the game.

To begin to uncover the difference between irrational status competition and competing to cooperate, we can start with the fact that status symbols, such as that alligator logo, elicit prosocial behaviors in all the cases we've just examined. The alligator logo does so because the brain interprets status symbols as clues to their bearer's value as a social partner. The evolutionary logic of social selection binds life success to the quality of our social partners—from those we choose as long-term romantic partners to friends and business allies. Because of the advantages such alliances provide, in all our social interactions we send signals to others about our underlying personal qualities that advertise our value as social partners. Although we do so mostly without any awareness of it, literally every aspect of our social behavior is such a signal. How we walk, how we talk, how we dress, how we style our hair, what we say, and what we don't say—all of these things send signals to other people about us. Likewise, our brain constantly interprets incoming signals from others to evaluate their value as social partners. This is at the core of the alligator effect. We give 25 percent more money to the high-status people in the Dictator Game, for example, because our brains evaluate them as potentially more valuable social partners. The money is our brain's way of signaling our interest in a social relationship with them. The fact that it does so even when there's no prospect of a relationship (because the game is played anonymously) is an important clue that social selection has shaped social reflexes that require no conscious deliberation or explicit strategizing.

Our examples so far have focused on luxury goods and their link to wealth. But it's important to note that wealth is just one form of social status, or sign of partner value. Many of our consumption motives are Pavlovian, or conditioned, Survival reflexes to signals of underlying partner quality. Even wealth is linked to partner value not just because it's a sign of resources, but because it can be a sign of intelligence, self-control, and other qualities. The routes to status are far more varied than wealth alone. The emergence of cool consumption depended on expanding these routes to status beyond wealth. What matters is that groups share common values about what confers status (social partner value) and use these to create cooperative alliances, whether the group is a hunter-gatherer tribe, a workplace, a circle of teetotaler friends, a motorcycle gang, a group of rebellious punk rockers, or a herd of hipsters. To understand what all these groups have in common, let's look at the *why* of consumption.

Who Do You Trust?

We're going to ask you to play a game. You'll play it anonymously with three other people over the Internet. Each of you will be given $20 to begin and may invest as much as you want into a group fund. All money that's invested will be doubled and then divided equally among the four of you. You'll keep whatever you make that round and start the next round with a fresh $20.

Let's suppose you'll play this for six rounds and everyone has complete information about how the group fund works. How much do you invest in the first round? If everyone invests their full allotment of $20, then you'd each earn a total of $40 each round. But here's the catch. Each dollar of your own money you invest only returns 50 cents to you, so earning money depends on how much others contribute. If you invest nothing and everyone else invests their $20, you'd make $30 from the game plus the $20 you kept, for a total of $50. Now, you may be wondering what other people are going to do: Will they be generous with their money, or will they be like you and take advantage of other people's generosity? If your group is a typical one, some players will invest substantially, but within a few rounds the contributions will decline, even those who were willing to cooperate in the beginning will stop investing, and the group fund will dwindle to zero.

This game, known as a Public Goods game, has been studied extensively, and it has been modified over the years as well. When the game involves two players, it's known as the Prisoner's Dilemma, which you may be familiar with—it refers to a problem facing two criminals who have been caught for a crime and are being held and interrogated in isolation from each other. The "dilemma" goes like this: Because the police don't have enough evidence to convict both people for the main crime, they plan to charge them each with a lesser crime that would result in a one-year sentence for each. So the police decide to offer each criminal a deal: If one criminal rats out his accomplice, he'll go free while his silent accomplice will get three years for the main charge. But if both criminals rat each other out, they'll each get two years. So if your accomplice rats you out, then it's better for you to rat him out as well so you don't get the full sentence while he goes free. If your accomplice stays silent, then it's also better for you to rat him out, since you'll go free and he'll serve the full sentence. Of course, your accomplice is likely thinking the same thing, so you both end up accusing one another and serving two years!

These dilemmas capture one of the greatest mysteries of social life: How is cooperation possible when self-interest undermines it? And these aren't just contrived academic problems—we can easily find examples of them in everyday life. Just look at climate-change issues. Dealing with climate change requires countries to reduce carbon dioxide emissions, but doing so may mean hurting their economy. So the leaders of each country are tempted to avoid reducing their own carbon dioxide emissions while pressing other countries to reduce theirs. Similarly, farming communities often share a much-needed resource, such as irrigation, but each farmer is tempted to overexploit the resource. The same happens with international fishing grounds. And if you listen to public radio but never donate during its pledge periods, you're free-riding on the generosity of others. It's the same with cheating on taxes.

The conflict between cooperation and self-interest is one of the most vexing problems of social life. Selfish incentives infect virtually every human cooperative enterprise you engage in. The search for romantic partners is fraught with worries that one's partner will not pull his or her weight, perhaps by being less emotionally engaged, putting self-interest first, not contributing fairly to the relationship in terms of commitment or care-taking responsibilities, or being unfaithful. Business relationships are likewise infected with these conflicts, from the office slacker to a business partner who may do less work than promised, may do substandard work, or may simply not follow through. At a more general level, selfish motives infect the ability of a society to create and maintain public goods: people cheat on taxes yet use public roads and other public goods and services. At a societal level, we refer to the capacity for and quality of cooperation as social capital, a sort of communal glue that's thought to contribute to economic prosperity and well-being.

The intrusion of self-interest on cooperative enterprises isn't necessarily a matter of malicious intent or explicit calculations of personal advantage—it doesn't have to be a conscious strategy, and it's usually not. Consider a scenario that is familiar to most of us. You're out to dinner with five friends. As you sit down, someone announces that you all should split the check equally when it comes time to pay. How do you think this will affect the total bill? The implicit reasoning is similar to the Public Goods game. Since the cost of your dinner is now being divided among six people, you'll pay for one-sixth of your dinner while the remaining five-sixths will be paid by your friends. So this provides an incentive to choose a more expensive meal than you may have originally planned. Of course, your friends think the same way. The result is that the average cost of each meal increases and you each end up paying more. In fact, a recent study found that the average cost of each meal jumps by 36 percent when dividing the bill equally![7] The pursuit of individual self-interest makes everyone worse off.

Evolutionary Dead Ends

If the problem of cooperation looks troubling from an economic perspective, it's downright depressing from an evolutionary one. Evolutionary thinkers have long recognized that group living promises many fitness advantages for individuals who can cooperate—hunting, food sharing, child rearing, defense, and so on. Yet the basic dynamic of natural selection is one of competition to survive and reproduce. Evolution by natural selection isn't just a struggle against nature: it is also a struggle against members of one's own species for reproductive success. The image of natural selection as making nature "red in tooth and claw" (a phrase from Tennyson that Richard Dawkins applied to evolution) is clear in the case of siblicide, when an animal kills its sibling. Consider the spotted hyena. Unlike most carnivores, spotted hyenas are born with their eyes open and a set of sharp teeth. They don't use these teeth for nourishment, since they depend on mother's milk for the first year or more of life. Instead, the pups, typically born in pairs, put these teeth to use by attacking one another in a fight for dominance. The loser will often starve to death, as its sibling denies it access to its mother's milk. This ruthless evolutionary logic has made the spotted hyena one of the most efficient survival machines on the planet. Lest we believe such logic doesn't apply to us, think of twins competing for nutrients in the womb, or the fact that murders involving adult brothers are more frequently committed by a younger brother in a power struggle with the older sibling.[8]

So we can think of natural selection involving other people as a variation of the Public Goods game. Instead of playing for money, the reward is fitness that gets cashed out at the end of the game as reproductive success—most simply, the number of offspring that make it into the next generation. (Considering that about 16 million men alive today are direct descendants of Genghis Khan as a result of his pillaging and siring hundreds of offspring, differential reproductive success can vary enormously!)[9] Just as an unconditional cooperator, or altruist, in the Public Goods game would continue to donate money to the common pool when others don't, an evolutionary altruist is someone who helps others at a cost to herself—a person who boosts someone else's fitness at the expense of her own.

Here's an evolutionary example: For our Stone Age ancestors, hunting big game carried the risk of getting trampled or gored. An altruistic ancestor would have been someone who risked his life to kill big game so that others could eat. Those who were more self-interested would have dilly-dallied during the hunt, letting the more altruistic take the risks. As you might guess, the evolutionary deck would have been stacked against the altruistic hunters, whose numbers would have slowly dwindled due to death by hunting.

As evolutionary theorists used computer simulations to study the interactions among altruists and selfish characters, they found that over generations, fewer and fewer altruists appeared. Just as the Public Goods game we invited you to play resulted in the decay of cooperation due to the invasion of self-interest, so too altruists lose out to the selfish across evolutionary time until altruists are no more. They are an evolutionary dead end. From the perspective of natural selection, nice guys really do seem to finish last. Dead last.

Does selfishness always get the upper hand? Is there no room for generosity, helping behavior, or cooperation in evolutionary theory? Surely it seems we are capable of unprecedented levels of cooperation—how else can we create and maintain the complex social structures we enjoy? Reconciling the apparently ruthless logic of natural selection with the intuition that we evolved a capacity for unprecedented forms of cooperation is the central mystery of social evolution. There are two main gambits that evolutionary theorists take in trying to solve this mystery. On the one hand, the influential work of W. D. Hamilton in the 1960s revealed how animals could be altruistic toward relatives. According to this view, a gene that helps generate altruistic behavior toward close relatives could be selected for by natural selection. That's because the close relative—the benefactor of the altruism—may likewise share the same altruistic gene. In this way, altruistic behavior toward close kin may evolve.[10] The most obvious human cases of altruism are parental sacrifices for their children. Since we know that the hormone oxytocin plays a central role in parent-child attachment, a gene that's linked to oxytocin function produces parental care, which leads to the most direct route for evolutionary success (namely, direct descent). There is no evolutionary mystery in why parents would be altruistic toward their child—it is because a gene that produces such behavior is likely to further the survival of its bearers' offspring and so persist from one generation to the next.

Hamilton, who had a background in economics, wondered whether a gene for altruism could evolve by means other than increasing direct fitness. At the same time, he had in mind a puzzle that Charles Darwin had struggled with: what appeared to be extreme cases of altruism in nature, such as the ways in which ants, termites, and some bees and wasps help rear the offspring of others but leave no offspring of their own. How could such "super-altruists" evolve? Darwin worried that such "super-altruists" would undermine his major theory, and even after the genetic basis of evolution had been described by Mendel, evolutionary theorists had struggled to explain what looked like a glaring exception to natural selection.

Hamilton worked out the costs and benefits of altruism in terms of indirect fitness. A gene that contributes altruistic behavior toward offspring is most likely to evolve, but it could still evolve by contributing to altruistic behavior toward other blood relatives. The reason for this is that since you share genes with blood relatives, you may share the gene for altruism as well. Hamilton knew that such a gene would be more likely shared with closer relatives than with distant ones. For example, while your daughter shares 50 percent of your genes, your niece shares 25 percent. So a gene that contributes to altruistic behavior toward a sibling may increase copies of itself to the population indirectly through a sibling's offspring, who may bear the gene. Hamilton produced a mathematical model of the costs and benefits of altruism and realized that a gene for altruism could evolve if the cost of altruism is less than the benefits it provides to blood relatives in proportion to their degree of genetic relatedness.

The evolutionary problem with altruism is, of course, that it tips the evolutionary scales against the altruist. Is there some way to make it less costly? Or is there even a way to reconcile it with self-interest? And how can altruism ever be in someone's evolutionary self-interest? Consider the following game. You don't know any personal details about the other player and they know nothing about you. You'll play the game with real money and pocket your winnings in cash. As in the Public Goods game, we are going to give you $20 at the start of each round and give you the option of either keeping it all or giving a portion of it to the other player. We'll triple whatever portion you give to the other player, so if you transfer $10 we'll make it $30. That player will then have the option to either keep all the money or to repay some of it back to you. Earnings from each round go into each player's piggy bank. A new round will begin after that with a new $20 for a total of ten rounds.

Just suppose you've decided to hand over the entire $20 to the other player. Let's pause the game there for a moment. This looks to be altruistic. You've bestowed a benefit on another at a cost to yourself. But of course the game isn't over. The other player needs to respond. Let's suppose they repay you nothing and pocket the entire $60. You're now handed a fresh $20, and the next round begins. Do you give them anything? Chances are, you're what's known as a conditional cooperator: you lead with a cooperative move, but retaliate when you meet up with a noncooperator. Alternatively, had the other player responded by repaying you $30, your first altruistic move would now be seen as having been in your self-interest, because cooperating had paid off. It pays for both of you to cooperate when you repeatedly interact and when there's a strategy available to you to punish noncooperators.

In the 1970s, Robert Trivers introduced the theory of reciprocal altruism, or direct reciprocity, as an evolutionarily plausible way for cooperation to emerge among non–genetically related individuals. The key idea, as the game above illustrates, is that an altruistic act will usually be reciprocated in the future, so both partners benefit from the interaction. There is, again, no requirement that the altruistic act be performed with an explicit expectation of repayment. The evolutionary logic behind this kind of cooperation is entirely consistent with the possibility that reciprocity is implemented through the psychology of friendship and its attendant feelings and sentiments.[11]

Human friendships are rooted in the evolutionary logic of reciprocity, and likely share some biological underpinnings with nonhuman friendships in other primates and other social animals, including elephants. Whenever we mention this fact to people, they bristle at the suggestion that their friendships stem from the evolutionary logic of reciprocity. They help their friends, they tell us, because they *like* them, not because they've calculated some sort of evolutionary cost-benefit analysis. Indeed, much work on the evolution of friendship distinguishes between exchange relationships (like the sort you have with a coworker you don't really socialize with outside of the office) and communal relationships (or close friendships). Exchange relationships more closely follow the evolutionary logic of tit-for-tat reciprocity and short-term repayment. In a communal relationship, you avoid repaying a friend's kindness immediately, since that looks as if you're treating the relationship like an exchange one. You also avoid explicitly keeping track of favors and obligations. But none of that runs counter to the fact that friendship is rooted in the evolutionary logic of reciprocity. For one thing, close friends do care about reciprocity, and friendships can easily dissolve if one friend feels betrayed by another's refusal to do an asked-for favor, or, conversely, feels asked to do too much.[12]

The bigger point is that answering the *why* question about friendships (Why do we have friends? Did friendship evolve?) is different from answering the *how* question about friendships (What psychological processes underlie our friendships?). Because friendships are important for survival, evolution shaped brain mechanisms and processes that create a psychology of friendship. At the psychological level, we really do care about our friends. There's no explicit calculation of costs and benefits. The same is true of our altruistic behavior, our consumption, and other social behaviors. Consider parenting, for example. Few parents decide to have children and to dedicate themselves to raising them with the explicit goal of transmitting their genes. The Survival pleasure machine within us takes care of that by making procreation and parenting rewarding. It structures our preferences, our desires, and the hedonic value related to parental care. Just as we don't eat simply to maintain homeostasis, but because we derive pleasure from food, so too the Survival system underlies genuine care and concern for children. The evolutionary answer to the *why* question of parenting is that caring for offspring is self-interested because doing so helps ensure our genes live on. But that doesn't mean parents don't genuinely care about their children. Instead, evolution shapes our psychology to deeply care about our children because our doing so perpetuates our genes. Our consumption is likewise shaped by social selection. But none of that means we deliberately pursue evolutionary goals in our consumption.

As appealing as the evolutionary logic of reciprocity is, it's not clear if it alone can account for how we became so intensely social and cooperative. Clearly, human social interactions involve reciprocity, but are we simply more "reciprocity-minded" than other social animals such as primates, dolphins, and elephants? Our sociality may share roots with the reciprocity systems of other animals, but something makes it radically different. To explain what made us so intensely social and cooperative—capable of creating extraordinarily complex societies in virtually every region of the world—we'll need to uncover another evolutionary force beyond natural selection.

Selection—Natural, and by Other Means

When we think about evolution, we tend to focus on natural selection, in which fitness stems from an organism's ability to succeed in an environment and against others for resources and survival. Darwin himself, however, knew that natural selection couldn't explain a lot of traits. Take ornamentation and animals' headgear, for example. The fact that such exotic plumage and antlers are typically found only on the male of a species flies in the face of natural selection. If they played an important role in survival, both sexes would have them. As Darwin struggled with these anomalies, he saw that mating competition was similar to other forms of competition that drove natural selection. He identified two forms of competition related to mating: intrasexual selection (males fight other males for dominance), and intersexual selection (members of one sex, typically male, compete to be chosen by a selective female). If individuals are choosy about their mates, it can start an evolutionary arms race over the traits they use to make their choice. So the trait has to be a good indicator of genetic fitness, an issue we'll return to when we explore the peacock's tail, the prototypical example of sexual selection.

Since Darwin, sexual selection has increasingly been proposed as an evolutionary force that has shaped a variety of human traits and behaviors, including intelligence, creativity, and consumption. But this connection leaves out an important component. The key aspect of sexual selection that drives evolutionary change is competition leading to mate choice. But what about other instances of selection that likewise involve human choice? Well, sexual selection is a particular type of a more general category known as social selection. Social selection refers to any form of partner choice that confers fitness. In our exploration of reciprocity, we encountered a prime example of partner choice—friendship. But we left out a critical detail. In the games of reciprocity that we asked you to think about, you weren't able to choose whom you played with. In fact, many of these games, and much of the evolutionary thinking based on them, operate by randomly pairing people together. But think about it for a moment. Do you randomly choose your friends? Do they randomly choose you? How do you decide whom to befriend? And what about circles of friends? Does the group make implicit decisions about who joins the group? What happens when these forms of partner choice become an ingredient in evolutionary thinking?

That's What Friends Are For

Just as sexual selection can drive runaway processes, such as the evolution of extreme traits, through competitive mate choice, so too social selection can drive runaway processes through the competitions that surround social partner choice. This broader form of social selection can have profound impacts on human evolution and social dynamics.[13] Indeed, much of the work in evolutionary theory has shifted from thinking about strategies to control randomly assigned partners to the process of partner choice and how it could underlie the evolution of social behaviors. We suspect partner choice was a game changer in social life and radically increased the amount of social interaction devoted to friendship, acquaintances, and peer-group choices.

Here's the basic idea. If we formed random friendships, alliances, and peer groups, a great deal of energy would have to go into controlling people's behavior, since the groups would contain all types of people. Social selection shifts the emphasis to investing energy in figuring out who are the best social partners to engage with in the first place. This is clear in the case of sexual selection, as we invest an enormous amount of energy in courtship, which some evolutionary psychologists liken to a moral obstacle course.[14] The reason we invest so much energy in choosing a long-term romantic partner is obvious: choosing the wrong person is a costly mistake. Similar pitfalls await poor decisions regarding our social partners—be it a close friend or a colleague at work.

In social selection for both romantic partners and social partners, there's an overlapping set of concerns: Will they genuinely commit to a long-term common endeavor? Or will they end up being something other than what they appear to be so you'll end up heartbroken, disappointed, or financially ruined? This problem of strategic uncertainty is so acute because we can't directly observe other people's internal traits. We can only guess their underlying qualities by observing their visible patterns of behavior—including their consumption. Of course, we have institutional solutions to some of this uncertainty in the form of legal contracts and recourses, but these are inefficient and cumbersome. Trust, commitment, and informal norms of cooperation—social capital—underlie efficient and productive societies, to say nothing of friendships and romantic partnerships.

People's choosiness about their social partners turns the problem of altruism on its head. Recall that the altruist is someone who performs actions that benefit others, at a cost to himself. This seems like a straightforward evolutionary dead end. But consider this. You're out walking one day in a busy part of town when you notice an elderly man walking slowly across an intersection. He's only halfway across when his light turns red. Just then, a woman rushes into the intersection and starts to wave at the oncoming cars to stop. She takes the elderly man by the arm and helps him to safety. Seeing this, chances are you form a good opinion of the woman. Forming these sorts of opinions comes so naturally to us that we may never have stopped to consider why we have them or the role they play in our social life.

To uncover the hidden logic of these opinions, recall the trust game we asked you to play earlier in this chapter. The researchers in Steve's lab had participants play a brain-imaging version of this game between players in Pasadena and players in Read Montague's lab in Houston. As people played, we could see brain signals emerge that were building predictions about the person they were playing with. Simply by looking at their brain activity, we could tell whether people were learning that the other person was trustworthy and cooperative or selfish and untrustworthy. In fact, as the game went on, we could predict how much they were going to offer the other person just by looking at their brain activity. In other words, their brains had formed opinions of the other player.

The ability to build opinions of other people is extremely important because it lets us accumulate information that will help us interact with them. The interesting thing about opinions is that once we have an opinion of someone, we start to rely more and more on it as a guide to interacting with them. We use our experience with someone to figure out what type of person they are—whether they are a kind and compassionate type or a duplicitous, scheming jerk. Once that's settled, we forgive kind people an occasional slip, while we may never trust someone we think is a jerk despite their often valiant attempts to regain our trust.

The opinion you form of the woman who helped that elderly man cross the street safely reveals an important twist in the mystery of human cooperation. We are able to form opinions of other people through direct interactions, but also by observing how they interact with others. This may sound like a simple matter, but it's likely a uniquely human ability, and one that is crucial for consumption and the rise of cool. That woman's altruism may have been risky to her—after all, she jumped out in front of oncoming cars—but when other people observe it, this "costly signaling," as evolutionary psychologists call it, increases her value as a social partner. In fact, this phenomenon starts as early as eighteen months of age, when toddlers help others selectively on the basis of the recipient's previous behavior. That is, they avoid helping children whom they've witnessed as "hinderers" rather than helpers.[15] Ultimately, the shared opinions we have of others coalesce into reputations. It's critical for life success to build a good reputation so others value us as social partners. If we use altruism as a criterion to choose our social partners, then once this process gets started it could lead to competitions to be more and more altruistic, what is referred to as competitive altruism.[16] So according to social selection, nice guys can finish first. And altruism isn't the only criterion we use. As we'll see, many traits and behaviors, including our consumption, could be shaped by social selection so long as they are used to make decisions about social partners.

Social selection solves the problem of altruism. More than that, though, its evolutionary logic reveals why our social life is so centered on trying to figure people out. Social selection explains why we care so deeply about our reputation. It explains why we gossip about others—both to learn about other people's reputations and to boost our own reputation at the expense of someone else's. And it explains why our social life is rife with typically unconscious displays of our value as a social partner, including our uniquely human use of consumption choices, from luxury logos to cool products, to convey to others who we are.

Networking Among the Hazda

A remarkable study of the social networks of the Hazda of Tanzania, a group of hunter-gatherers, delved into the question of whether "choosy" partner selection played a role in the evolution of cooperation. Surviving hunter-gatherer groups are an important source of evidence regarding the conditions and pressures that may have played a role in our social evolution, and we will meet various groups in later chapters. The researchers ran a variation of a Public Goods game that was suitable for the Hazda. Since the Hazda have no currency, the researchers used honey sticks, which are highly valued by the group. They gave about two hundred Hazda four honey sticks each and told them they could keep them all, but that each stick they contributed to an anonymous pile would be tripled and then redistributed. The researchers then examined whether there was any relationship between a member's cooperation and the nature of the social network, the network of friendships, that person belonged to.

By examining the pattern of gift giving and campmate networks, the researchers found that Hazda cooperators tended to be friends with other cooperators. This "clustering" allows cooperators to benefit from each other. Noncooperators clustered with other noncooperators (perhaps because the cooperators excluded them from their networks). Like modern social networks (including online communities as well, such as Facebook), Hazda social networks aren't random—people are choosy when it comes to their social partners. This study and others like it suggest that a great deal of our social behavior relates to the pressures of social selection. As we'll discover, our social life, including our consumption behavior, is permeated by these pressures.

Social Signals

In recent years, costly signaling theory has been applied to a wide spectrum of both human and animal behavior.[17] The basic idea is that a range of behaviors and displays signal important information about an individual, from economic resources and social standing to character, intelligence, physical skill, and cultural knowledge. By using these signals, we communicate, typically unconsciously, to others what type of person we are and what social, economic, or political benefits can be gained by interacting with us. Others, in turn, use these signals to decide how, or even whether, to engage in cooperative enterprises with us—whether we are trustworthy and would make a good partner, friend, ally, or business associate. It's no accident that across all types of interdependent relationships, from athletic teams to friendships and families, trustworthiness is cited as the most important single characteristic.[18]

A successful signaler is also the recipient of many benefits, some of which may include increased social status, lucrative trading partnerships, allies in times of conflict, supporters in times of need, and the selection of reliable long-term mates, all of which can be turned into resources that influence a person's short- and long-term fitness.

None of this signaling—not the sending, receiving, or evaluating—needs to occur at the conscious level. What matters is that signals trigger certain behaviors. There is no need, for example, for a peahen to recognize that she prefers peacocks with larger tails for her mate choice to be influenced by tail size. These preferences have been shaped by evolutionary forces that influence circuits in the peahen's brain to respond to prospective mates on the basis of their plumage. The peahen is in no way aware of what underlies her preferences, just as peacocks are not consciously aware of any link between their plumage and their attractiveness as mates.

We suspect that many of the signals we use and respond to operate below our awareness. What's more, the deceptive use of these signals may also be unconscious, for the simple reason that we may often lack accurate knowledge of our own underlying qualities. The brute evolutionary fact of the matter is that our own self-perception is fraught with many biases that make self-knowledge problematic at best. Most people, for example, evaluate themselves more favorably than others do on a broad range of traits, qualities, and actions.[19] This tendency toward self-enhancement—from self-evaluations in online dating profiles to rating one's own driving skills—suggests, along with other self-biases, that having a positive self-concept is more important than having an accurate one. For this reason, we may unknowingly signal ourselves to be better than we really are, even without any conscious intention to deceive. Much of our social life, then, is a complex blend of conscious and unconscious signaling aimed both at advertising ourselves to others and at figuring out what kind of people others are as we weigh existing alliances and friendships or decide to move forward with new relationships. These signals range from natural ones, such as facial features, to elaborate behaviors and rituals—and to the goods we consume. They constitute a complex and typically nonverbal communication system that makes our everyday social exchanges possible. We are awash in these signals: How we style our hair, hold our fork, greet our colleagues, tell a joke, and give gifts are all elements of social signaling.

There is another facet of this signaling worth noting. Signaling is tied to creating and maintaining cooperative relationships, and this

adaptive role answers the *why* question about our signaling behaviors in terms of the evolutionary logic that shaped them. As for
the *how*, evolutionarily successful behaviors are typically built into
our brain's reward systems. We're not conscious social-value calcula-
tors (although we certainly can be sometimes, just as we sometimes
eat to maintain homeostasis or even have sex specifically to procre-
ate). Rather, at the psychological level, signaling behavior is itself
intrinsically rewarding and can happen in the absence of any explicit desire to create new relationships. In other words, signaling
is not an explicit strategy we consciously pursue only when we believe it is advantageous. It is a ubiquitous feature of our behavior
because it has a direct route to our brain's reward circuitry through
our sociometer.

Positive social feedback is intrinsically rewarding. Even college
students rank it above food and sex in their reward priorities. The
mere anticipation of positive social feedback is rewarding as well.
Just driving an expensive sports car around town, knowing he's being seen at the wheel, is enough to raise a man's testosterone level—a
sign of the increased sense of status registering on his sociometer.
As we'll discover, consumer goods are efficient social signals, particularly in complex societies in which there is both a great deal of
heterogeneity and frequent interactions with strangers that require
rapid evaluations. Beyond our need to make decisions about others,
complex societies also create unique exposure or audience opportunities for signaling—that is, extensive occasions to obtain perceived
social feedback from strangers. Before we explore signaling and
consumerism in more detail, let's explore the diversity of signals we
all transmit as we go about our daily lives.

Natural Signals

Consider the human face. Even your briefest glance at a face sets off
a cascade of unconscious processes in your brain. One remarkable
example of this is a variation on the trust game in which you are
now presented with a picture of the face of the person you will play
with. Does your brain rapidly evaluate that face, making judgments

that will influence your behavior? Of course. One robust cue to trustworthiness is facial attractiveness, and we tend to trust attractive people more in the trust game.[20] In addition, we tend to trust faces that resemble our own.[21] For experiments like these, we can use computer face-morphing software to transform the image of a face to make it look more or less like the experimental subject's own.

Most people don't consciously realize that self-resemblance has an effect on their trust decisions, that self-resemblance is a cue of genetic relatedness, and that trusting genetic relatives more than nonrelatives follows the evolutionary logic of kin selection. We are all more likely aware of basing our trustworthiness judgments on attractiveness, though we might deny that it's a good reason. One feature that we are probably unaware of is facial-width ratio (the width of a person's face relative to its length). In an intriguing experiment, Michael Stirrat and David Perrett of the University of St. Andrews in Scotland found that the trusting behavior of both men and women varied with the facial-width ratio of a male face presented to them as that of the man they were playing a trust game with.[22] As facial width increased, players trusted the other player less, even after taking facial attractiveness into account. What is intriguing about this study is that subjects were right to be suspicious of wide faces: such people are in fact more likely to exploit trusting behavior, making facial width a valid cue of trustworthiness.

As it turns out, facial width is influenced by testosterone effects during adolescence, making this feature a reliable signal of testosterone-related traits, including male aggression. A Canadian study found that the facial width of professional hockey players was a good predictor of how many penalty minutes those players received during a season.[23] In fact, people can accurately judge a man's propensity for aggression even when they see his picture for a mere one-twentieth of a second.[24] Since facial width is a reliable signal of aggressiveness, these fast and automatic reactions to facial cues may have evolved over time without any awareness of what specific cues we were reacting to. Indeed, it is easy to appreciate the evolutionary importance of being able to quickly and accurately judge the aggressiveness and trustworthiness of others based on facial cues.

A number of brain-imaging studies have also revealed that the brain quickly and automatically categorizes faces according to their perceived trustworthiness.[25] In one experiment, subjects were told they were participating in a face-memory study, so they were not asked to explicitly evaluate faces for trustworthiness.[26] Instead, they were told to decide if a face they were shown appeared in a subsequent series of faces, each presented for just a second. Only after fMRI scanning during this task were the participants asked to rate the faces again on a scale of trustworthiness. Researchers then examined the activity in the amygdala (as you'll recall, it's involved in emotional processing and has been linked to fear and threat detection) during the memory task. They found that the higher a person rated a face's untrustworthiness, the greater the activity had been in their amygdala when they briefly viewed that face with no conscious thought of trust. While the participants were engaged in a distracting neutral task, their amygdala had been busy forming opinions of other people below the level of their awareness.

These results aren't just found in the lab. A study of more than five thousand loans from the peer-to-peer online lending site Prosper revealed a strong link between faces and real-world trust. A few dozen people who weren't involved in the loans were first asked to rate the borrowers' faces in terms of trustworthiness. Then, examining the loans themselves, researchers found that the faces rated as more trustworthy were not only more likely to have been offered a loan, but were offered one at a lower interest rate than borrowers whose picture was rated as less trustworthy.[27] While a face may seem like a superficial basis for making loan decisions, it turns out that people with more trustworthy faces did in fact have lower default rates.

These results reveal a number of important facts about how our brains assign value to natural signals, such as faces; how these evaluations structure our preferences; and how these affect our decision making. Faces signal values that our brains respond to, and they use these values to guide our behavior without our being aware that this process is going on. In fact, not only do our explicit attitudes frequently disagree with our implicit attitudes, but we often vehemently deny that we make decisions on the basis of such things as how people look. Of course, from the perspective of the brain it doesn't matter that you are unaware of these values, and it may actually facilitate your behavior to keep these motivations implicit, in the same way that deception works at many other levels. The core view of our understanding of human economic life is that we are awash in a sea of these implicit signals, and that much of the value in our economy doesn't reside merely in "things" but in their capacity to act as social signals that our brain rapidly decodes.

Behavioral Signals

A male teenage driver pulls up to a McDonald's drive-thru and orders an ice cream cone, secretly recording the transaction on his cell phone. He pays and pulls up to the pick-up window where the McDonald's employee hands him the ice cream cone. Instead of reaching out for the cone, the young man grabs the ice cream. Holding the ice cream in his hand, he starts to eat the cone first. The McDonald's employee looks befuddled as the teen drives off. A You-Tube video documenting the aforementioned episode has been viewed more than 10 million times.[28] It was made a few years ago when teenagers were in the midst of a craze known as "coning," and it's one of countless videos like it. Of course, what made coning funny was the reaction of the McDonald's employees to deviant ice cream handling and consuming. In one instance, a teenager takes the cone normally but smushes the ice cream against his face inside his car. When he returns for another ice cream cone, the employee refuses to hand it over. A minute later an annoyed manager appears at the window, notifies the teenager that she'll call the police, and tells him that the drive-thru is for "normal" people.[29] The teenager tries to object that he should be able to eat the cone any way he wants, but the manager is having none of it and warns him again that he will be arrested if he doesn't leave.

The reaction of the McDonald's employees highlights the fact that virtually every facet of our social behavior involves social norms—even ones as seemingly mundane as how to eat an ice cream cone. While coning may seem like just a silly prank, it fits into a tradition of testing the pervasiveness of social norms by violating them and observing people's response. In fact, the sociologist Harold Garfinkel developed what are known as breaching experiments to investigate how people respond to social norm violations. For example, he had his students act as boarders in their own homes when they returned to visit for the holidays. When the students behaved with the formality of a boarder—for instance, asking to use the bathroom and being particularly polite—their parents often reacted with anger, or were worried that their child had had some sort of mental breakdown. Other experiments have involved norm violations such as riding an elevator facing the back wall, standing too close to strangers, or trying to negotiate bus fares. Even these minor norm violations were met with serious objections from onlookers. Of course, this hasn't escaped the notice of comedians, from Seinfeld to Sacha Baron Cohen (creator of Borat), who routinely premise their humor on social norm violations.

Social norms are the typically unspoken rules and beliefs that regulate our social behavior and shape our attitudes about how others should behave. The philosopher and psychologist Cristina Bicchieri refers to social norms as creating a kind of grammar of social interactions, a system that prescribes appropriate behaviors and helps to create shared social identities, group conformity, cohesiveness, and social coordination.[30] One of the most important aspects of social norms for us is the fact that they are *collective* beliefs about appropriate behavior: they prescribe behaviors, and create expectations and attitudes about behaviors, that are shared by a group and define boundaries among groups. Shared social norms are salient in particular at the level of our social identity, our self-descriptions that draw on our group membership.

Norms no doubt play an essential role in coordinating social behavior—it's useful to have norms telling you which side of the street to drive on, for example. However, norms can't be explained purely in these terms. Norms possess a number of important properties that make them excellent signaling mechanisms, which is a reason why they've become so elaborate. Facial features are powerful social signals because they are readily accessible to others and can be processed quickly. Despite these strengths, however, faces are relatively inflexible, and are most likely limited to signaling basic information regarding underlying traits. Behaviors, in contrast, ranging from facial gestures to complex rituals, offer a wider repertoire of social signals. Virtually every human culture has taken advantage of the signaling power of behavior by creating complex social norms around everyday family, social, and business behaviors.

Following social norms is a strong prosocial signal that reveals your willingness to observe the standards that govern your society. In fact, looking at signaling and the law, the legal scholar Eric Posner examined how much of our social order is dependent on people's willingness to conform to social norms as a signal of their desirability as social partners.[31] The fact that social norms are often more complex than they need be—the latest Emily Post etiquette book comes in at a backbreaking 736 pages—reveals that these norms require a substantial investment of time to learn, indicating that they are costly signals. From mundane to complex rituals, norm-following behavior pervades our social lives. While norms such as what side of the road to drive on help coordinate society, many others are powerful clues about social partners. Importantly, material goods and lifestyle activities amplify behaviors by giving them deeper meaning: there's a big symbolic difference between riding a Harley and a Ducati, despite the fact that on the surface level it's the same behavior. A favorite breaching experiment among students is to mismatch behavior and goods, like working out at the gym in a business suit, which grabs attention.

Signaling Goods

Natural signals, such as faces, and behaviors, such as dining etiquette, offer a wide array of signals. But behaviors, like faces, have some serious limitations. We've already seen that while faces are highly visible social signals, they have limited flexibility, even with entire industries devoted to manipulating their appearance, from cosmetics to plastic surgery. Behaviors, by contrast, are supremely flexible. But many behaviors require relatively long periods of time to observe and evaluate. In some cases, such as courtship, people often take months or even years of observing and evaluating a romantic partner before they feel confident that they are making the right choice. Our ability to use goods to signal our value as social partners was a remarkable innovation and was likely a key to creating complex societies. Like faces, signaling goods (from clothes to motorcycles) are socially visible and can be assessed rapidly by observers. They can also be broadcast to a large group simultaneously. Unlike faces, however, goods provide numerous signaling possibilities, similar to behaviors in this sense. Indeed, goods and behaviors are often coordinated, and goods can be used to predict how their owners are likely to act. To give you a sense of the diversity of signals we can send, we've arranged the traits that people consider most important in others in Table 1 below.[32] We can use products to signal all of these, from thick-rimmed glasses to look smart to funny, whimsical accessories.

Partner Trait Category	Trait Attributes
Trustworthiness	Honest, loyal, sincere, dependable
Cooperativeness	Sharing, fair, just, supportive
Agreeableness	Kind, interpersonally warm
Extraversion	Outgoing, sociable, friendly, funny
Conscientiousness	Organized
Emotional Stability	Calm, happy
Open-mindedness	Open-minded, creative, unconventional
Intelligence	Intelligent, rational
Assertiveness	Assertive
Attractiveness	Physically attractive
Physical Health	Healthy

TABLE 1. Desirable Partner Traits and their corresponding attributes. Consumer goods signal an array of these attributes.

Recall the woman in the Lacoste sweater. In a fraction of a second, the people she encountered in the mall perceived that alligator logo, sending off a cascade of brain processes that first recognized the logo by matching it to their memories, then recalled information they associated with that logo, formed an implicit opinion of the woman based on those associations, and then generated the motor commands that resulted in the utterance indicating a willingness to partake in the survey. Little of this was conscious. Indeed, the unconscious processing of signaling goods can dramatically impact the opinions we form of other people.

Suppose, for example, that you're watching a video of a man being interviewed for a job. Your task is to evaluate his suitability for the position and to suggest an appropriate hourly wage. Would an alligator logo on his shirt affect your decisions? In an experiment, some participants saw a version of the video in which the man (who was really part of the experiment) wore an unbranded white shirt. Others saw a version in which the same man wore the identical shirt except that it now had a luxury-brand or a non-luxury-brand logo on the front. Everything else about the videos was identical. The participants who watched the Lacoste shirt version did in fact rate the man as more suitable for the job and suggested a higher hourly wage than did the participants who saw the other version.

The Lacoste shirt we've been asking you to think about is one of the oldest examples of branded clothing (with a brand logo appearing on the outside of the clothing), originating in 1933. Yet the use of symbolic goods is not only as old as humanity, it's what anthropologists consider to be the sign of the emergence of humanity itself. In other words, it's what distinguishes humans from other animals. The earliest known use of symbolic goods traces back to around seventy thousand years ago, when some humans wore shell necklaces.[33] We can only speculate on what those shell beads meant to their bearers, but their significance lies in the fact that they involved a symbolic signaling of personal identity to others. In the early exchange networks humans were forming then, they may have been ways to signal to other groups that their bearer was a member of a friendly group interested in exchange. As long as we've been around, we have been using things to signal who we are to others to facilitate social exchange.

Seventy thousand years ago, a shell necklace on an approaching stranger may have been enough to distinguish a friend from an enemy. That necklace—among the very earliest signs of humanity— was an archetypal example of a good that signaled trustworthiness and cooperation; thus, it illustrates the logic of social selection. Indeed, it's not so different from the goods cool consumers are using seventy thousand years later to signal their value as social partners. As social life becomes more complex, social uncertainty grows too, which amplifies the need for visible and rapidly evaluated signals. We probably encounter more strangers in a day than even our not-too-distant ancestors did in a lifetime.

Social mobility also decreases the duration of many direct interactions. Kerwin Charles and his colleagues found that goods signaling in fact increases when social uncertainty is high. One way that social uncertainty can be high is if you belong to a social group that strangers may think is a relatively poor one. In this case, you conspicuously consume to show strangers you aren't poor. To test this, the researchers compared what percentage of their income African Americans, Latinos, and whites of comparable demographics (family size, income, and so on) spent on signaling goods, including cars, clothing, and jewelry.[34] They found that on average African Americans and Latinos spent up to 30 percent more of their income on conspicuous consumption. There were no spending differences in items such as home furnishings, which are less visible signals. These effects were greatest in geographic areas that were relatively poor—where one was more likely to be considered poor by strangers. In fact, when the researchers examined regions that contained relatively poor white Americans, they found that they too spent considerably more on signaling goods. In other words, when people are surrounded by relatively affluent peer groups, they can spend less on signaling goods because people are more likely to assume they too are affluent. These effects were particularly strong among younger people. This, too, fits the logic of social selection, as signaling is expected to be higher among younger people, particularly as they enter the workforce.

Putting the Social Back in Social Consumption

The stock views of consumption we considered earlier in this chapter focus on the individual only—how consuming satisfies individual desires and needs, from the need for distinction to the need for status. Consuming, in these views, is about standing out from the crowd, creating social distance. When these perspectives highlight the symbolic nature of goods, it's typically in this context, as ways individuals mark off distinctions from others. To put it starkly, these views picture consumption as building fences, not bridges. Viewing consumption from the perspective of social selection reveals why these stock views distort the nature of consumption and why we'll need to go beyond them to explain the rise of cool consumption. Our starting point lies in recognizing that consumption is tied to social cooperation.

Indeed, perhaps the deepest insight to come from brain science over the last few decades has been that the brain structures critically evolved in our social consumption—especially the MPFC—are ones that social selection shaped to make our species social in ways unlike any other. The complexity of the cooperative alliances we create, from intimate relationships to citizenship of countries, make us intensely "groupish." As a result, much of our consumption rests on basic affiliative impulses of social bonding. The sociometer itself evolved as part of the human neural machinery that motivates us to seek the rewards of social bonds. The upshot is that our consumption is typically more about conformity and social belonging than individual standing out.

Researchers increasingly focus on social consumption, variously describing consumption groups as consumer tribes, brand communities, or consumption microcultures. This reflects a shift from viewing consumption as the act of isolated, competing individuals to viewing it as a social enterprise. The essential element of consumption as a social enterprise is that social groups share consumption norms and members gain esteem through their inclusion in these social groups. Consumer tribes and communities tend to form around specific brands, such as Harley-Davidson and REI. Others, particularly among youth, tend to form around lifestyles represented by musical preferences, from punk to hip-hop. Because so-cial selection makes us choosy about our social partners, group membership itself bestows self-esteem, particularly as we create and participate in the social identity the group fosters. Belonging to the group is rewarding because the sociometer represents more than just your individual self-esteem.

There are at least three levels of self-representation, each linked to the brain's reward structures. These are our individual self, our relational self, and our collective self. Think of your individual self as the collection of traits that make you a unique person. Your rela-tional self is the way you think about yourself in relation to signifi-cant others, as a spouse, friend, or a parent, or even in work relationships, the way a lawyer might think about his identity in terms of his relationships with clients. Your collective self is the way you think about yourself in terms of larger groups that you identify with, such as your identity as a member of an ethnic or religious group, or as a political liberal, conservative, or libertarian, or even as the fan of a particular sports team. Our relational and collective selves often involve "consumer tribes" that revolve around the kind of car we drive, our recreational activities, and so on.

These differences in self-identity have intriguing implications for how we think about status. The status concerns traditionally thought to underlie conspicuous consumption (we examine status concerns in detail in the next chapter) are highly individualistic, in that status is traditionally defined as the influence or prominence individuals have in a social group. You can think of individual status as the rungs on a ladder, where only one person can occupy a rung. When one moves up, someone else has to move down. This is the status of the individual self. In contrast, consider the sense of pride you get from belonging to certain groups. If you're patriotic, for example, you may feel a swelling of pride when you listen to the national an-them at a public event. Social psychologists refer to this feeling as collective self-esteem, and they measure its strength by asking peo-ple whether they feel the groups they belong to reflect their values, are respected by others, and so on. Collective self-esteem, unlike individual self-esteem, isn't a limited resource. You don't feel less proud, for example, just because more people share your collective self-esteem. In fact, collective self-esteem often grows the more people share in it. We feel we gain respect from our collective ties, which differs from individual status.

As the psychologist Cameron Anderson and his colleagues have shown, status and respect are separable.[35] Individuals desire high status in the abstract, but don't always seek it in real-life contexts. Instead, the level of status we seek depends on our self-perceived value to the group. Individuals who believe they possess the abilities to contribute to the group's success pursue higher status, while those who believe they lack such abilities opt for lower status. Even when individuals opt for lower status, they still desire high levels of respect when asked how much being held in esteem matters. We believe this distinction between status and respect underlies motivations at different levels of the self and that our consumption is as much about the desire for respect as it is about status as an individual. In fact, when our collective self is salient, we often act in ways that increase the group's collective self-esteem at the expense of our individual self.

Recall the Public Goods game we asked you to play earlier in this chapter. We asked you to play it as an individual—you didn't know any of the other players. But if we had first sorted the players into teams, the outcome would have been very different. We could have made teams by a criterion as simple as dividing people into two groups on the basis of their preference for one of two paintings (say, a Rembrandt vs. a Picasso). But as soon as you belong to one of the groups, you start to show favoritism for your own group. Soon, you're playing the game to maximize your team's winnings even if it means less money for you. Playing the Public Goods game in painting-based teams is an example of what social psychologists refer to as the minimal group paradigm. A long history of experiments reveals that even when groups are temporary and arbitrary, in short order members show a strong preference for their own group, a phenomenon that underlies the biases that pervade our social lives.[36] Of course, it doesn't require a trip to the lab to confirm this. Just go to a sporting event.

Cool consumption can't be explained without broadening our view of consumption to see it as a social enterprise that includes collective self-esteem. Likewise, it requires seeing the fences and bridges that people build through consumption in light of the in-group and out-group dynamics that are driven by how our brain links social rewards to our collective selves. To uncover these dynamics, we'll need to see how consumption signals morphed from those signaling wealth to those signaling a wider variety of social traits. In the next chapter, we'll explore how signaling interacts with these social and cultural forces more generally to drive the proliferation of consumer tribes and the rise of cool consumption.

5

STATUS SEEKING AND THE REBEL INSTINCT

Sturgis is a sleepy little town of about seven thousand in the sparsely populated state of South Dakota. To most travelers, it's known, if at all, as one of the few towns along Interstate 90 where you can stop and get gas between Rapid City and Spearfish. Since 1938, however, for a week in August the otherwise quiet streets of Sturgis are shattered by the deafening roar of hundreds of thousands of motorcycles when the city plays host to the Sturgis Motorcycle Rally. For as far as the eye can see, rows and rows of motorcycles line the city streets; the air is filled with the scents of beer, leather, and exhaust. The rally now draws about half a million motorcycle enthusiasts of all stripes.

Sturgis is also the unlikely research site for two marketing professors, John Schouten and James McAlexander. For the last few decades, they have traveled to rallies including the one in Sturgis and have ridden countless miles alongside Harley-Davidson owners as ethnographers in what may be the longest and most detailed research project on the evolution of a consumer subculture.[1] When Schouten and McAlexander began their study of the Harley-Davidson subculture in the early 1990s, Harley-Davidson sold about 70,000 motorcycles annually. At the time, the Harley-Davidson subculture was a relatively homogeneous and hierarchical one, centered on the norms of personal freedom, patriotism, and machismo. Even in the early years of the study, however, the traditional outlaw image of Harley-Davidson bikers belied diverse subgroups, some of which were chapters of HOG, the Harley Owners Group that Harley-Davidson began in 1983 as a company-sponsored riding club. One chapter, for example, serves as a support group for recovering addicts and alcoholics, another for Vietnam veterans, and one chapter is even a born-again Christian club that huddles around a motorcycle radio on Sunday mornings to listen to religious services.

A lot has changed in the world of Harley-Davidson owners over the last thirty years. By 2005, Harley-Davidson's annual sales had shot up to more than 325,000 bikes. And according to Schouten and McAlexander, "In these intervening years we have witnessed the death of the relatively monolithic subculture of consumption that we first encountered. In its place we have observed the emergence of something larger and richer, something we are more comfortable thinking about as a complex brand community or a mosaic of microcultures."[2] In particular, in place of the predominantly white male baby-boom population of the early 1990s, now more women, Gen-Xers, and other ethnic groups participate, which in turn has led to a broader range of lifestyles among riders. Members of the Harley-Davidson groups find a great deal of meaning and significance as they become more involved in those lifestyles, a process Schouten and McAlexander describe as identity transformation and self-reinvention. These consumer subcultures contain quasi-religious (sometimes literally religious) and ritualistic elements, and strong feelings of community identity that members describe as a brotherhood of shared belief and experience. This process of self-invention and reinvention and social connectedness through consumption patterns is in stark contrast to the disapproving image of consumerism as shallow and solipsistic, and points to the affiliative logic of social selection.

Why did the Harley-Davidson consumer culture evolve from a hierarchical to a pluralistic one, a "mosaic of microcultures"? For that matter, why are similar transitions from hierarchical culture to pluralistic microcultures a pervasive theme in the recent history of consumerism? In fact, many consumer culture studies over the last few decades point out the proliferation of such microcultures, the proliferation of lifestyles, and related trends toward consumer diversification.[3] Indeed, the growth and diversification of consumer culture itself coincides with a more general trajectory of many societies along similar lines. The political scientist Ronald Inglehart has studied large-scale social change since the 1970s. Since 1981, he has been the director of the World Values Survey, a massive series of national surveys that now poll people in nearly one hundred societies representing 90 percent of the world population.[4]

Inglehart found a titanic shift in values across generations, starting among the postwar generation who came of age in the late 1960s and the 1970s. Older generations held what Inglehart describes as values that were shaped by the economic realities of growing up in a world where material sustenance and physical security were scarce or uncertain. As a result, people prioritized the materialist values of economic and physical security. Postwar economies, however, brought prosperity and increasing abundance. With these new realities, young people came to hold a set of "post-scarcity" economic, political, and cultural values. These emphasize autonomy, self-esteem, self-expression, aesthetic and intellectual satisfaction, rejection of authority, and tolerance for lifestyle diversity and individual self-expression, which we examine in more detail in chapter 8.[5] As the Princeton University historian Daniel Rodgers notes in *Age of Fracture*, an influential intellectual history of the United States in the last quarter of the twentieth century, social fracture characterized this period. That is, large social movements and collectives splintered into increasingly fragmented ones. Sociologists, economists, and political scientists also shifted their explanations of social change from analyses emphasizing large-scale social structures to smaller-scale explanations focusing on individuals, a shift of perspective perhaps most famously captured by Margaret Thatcher's remark that there is no such thing as society.[6] Political scientists and historians such as Inglehart and Rodgers trace the rise of an increasingly diverse, fragmented, and pluralistic society, but what forces within us helped drive these transitions, particularly in the changing consumer culture?[7]

In this chapter and the next, we'll put the neural machinery of social signaling into the broader context of cultural biology to explore how human status motives interacted with social and cultural forces to first create and then diversify status systems. The structure of these forces is so basic that we share essential elements of them with our closest genetic relative, the chimpanzee. Chimps live in hierarchical groups, strive for status, and sometimes band together to rebel against dominant chimpanzees. Indeed, for most of history, it appears that humans likewise lived in hierarchical status systems. We'll see that hierarchical societies, such as chiefdoms and the city-state, emerged alongside scarce, defensible resources with striking rapidity and frequency. This is because we too possess a "status instinct," which disposes us to seek status and to compete with others for it, a motive rooted in the logic of social selection, which fuels the formation of status hierarchies.

Unlike our ape relatives, however, we humans are also motivated by the status instinct to enforce status differences through social institutions that create systems of social stratification. Social institutions emerged in our world that integrated cultural ideas and social roles into a social order that codified hierarchical status. At the extreme, this created status systems such as the Hindu castes of India, where society was segmented according to hundreds of rigidly defined groups, occupations corresponding to those castes, physical segregation among castes, and strong social norms regulating interactions (such as marriage) among social groups. In such a world, individuals and groups are not merely different: one's place in the hierarchy corresponds to one's social status and determines one's access to valued resources, including income, prestige, and prominence.

The status instinct also motivates the human thirst for political ideology that tries to justify social stratification, perhaps rooted in what primatologists call the "conservative coalitions" that chimpanzees create to support their social order. Here are the primordial roots of conservative political sentiments, which include aversion to change, dislike of uncertainty, and opposition to equality.[8] These sentiments produce ideologies—such as beliefs about gender roles, purported racial differences, and meritocracy—that attempt to justify hierarchical social order. In fact, by investigating a personality trait called social dominance orientation, researchers have uncovered the genetic and evolutionary origins of political orientations, which help to explain why some embrace more conservative political ideologies than others.[9] Far from a modern invention, then, political orientation is rooted in the ancient logic of the quest for status. Together, these forces create a ruling elite, the status quo establishment, a dominant group that controls access to resources and status by subordinating others.

Of course, many formal institutional systems (such as the military, businesses, and universities) also incorporate a hierarchical status system. Status may be marked by title, dress, or turf—as in the relative size of one's office. Throughout human history, and even in prehistory, hierarchical status systems seem to have emerged whenever there were scarce, defensible resources for people to inherit.[10] Many such stratified societies also regulate consumption through sumptuary laws, which enforce social hierarchies by regulating consumption according to social rank. Essentially, for as long as there have been legal codes, these codes included restrictions on consumption, particularly regulating dress according to social class. These sumptuary laws helped to systematize status in well-defined hierarchies and quelled the proliferation of alternative status systems—two hallmarks of traditional societies.

Hierarchical status systems and the emergence of the status quo establishment create what we'll refer to as the Status Dilemma, the "zero-sum" status contest that forms the core of many critiques of consumerism.[11] These typically focus on well-ordered status hierarchies, ranking systems in which each individual is assigned a status rank.[12] Picture a pyramid with fewer and fewer positions the higher you go. The only way to ascend is to knock someone above you out of their spot. Thorstein Veblen's theory of conspicuous consumption is the prototypical example, and many contemporary critiques retain the same basic logic.[13] Because high-status rank is so limited in a clearly ordered social hierarchy, only a few people can have it. The overwhelming majority are destined to be frustrated and unhappy, in the anti-consumerist view.

Consumers are trapped by a second element of this dilemma: the psychological motive of emulation, copying the consumption patterns of those above to gain rank. Those belonging to lower status groups emulate those of higher status and seek to raise their own status through emulation. Emulation works because in a hierarchical system people recognize and agree where goods fit along this status gradient, and so one's possessions transparently signal status. Here begins the zero-sum game: high-status consumers introduce a new taste, people of lower status emulate them, higher-status people then abandon the taste because it's become popular among the lower status, and the next cycle of imitation-abandonment begins. Consumerism's critics often invoke the Easterlin paradox here to claim that this cycle of consumption doesn't make anyone happy. Like the Prisoner's Dilemma, in which the individual pursuit of self-interest makes everyone worse off, the Status Dilemma stems from the idea that our individual pursuit of status concerns and emulation makes everyone worse off.

But these critiques are fatally incomplete, because they leave out the flip side of the status instinct. As with our chimpanzee cousins, because the status instinct leads to rivalry, it creates an opposing force. We call this the "rebel instinct," a deeply rooted emotional aversion to being subordinated. In chimpanzees the rebel instinct can sometimes lead to what primatologists call "revolutionary coalitions," and even to deadly coups.[14] Similarly, from the quarrels over power in small bands of hunter-gatherers to modern revolutions, the rebel instinct fuels our anger, frustration, and resentment when others try to dominate us—that is, when we grasp what psychologists refer to as a sense of relative deprivation compared to the ruling elite.[15] Although consumption critiques still rely on picturing consumption as a zero-sum contest for status in a hierarchical society, a crucial change in the human rebel instinct transformed social hierarchy. The chimpanzee rebel instinct only shuffles the status order. Human rebellion can reject the status quo and create alternative status systems. Our capacity to create lifestyle subcultures and countercultures depends on our status and rebel instincts, and together these created the dynamic of oppositional cool consumption.

To examine how these motives played a role in the evolution of consumer culture, we'll draw on a key idea from evolutionary biology. One evolutionary response to competition for limited resources is diversification, which diffuses competition. Although we often think of evolution as primarily a "culling" process, one of the most remarkable facts of the natural world is the sheer number and diversity of species. Pitted against selective pressures is an equally extraordinary process of creation, diversification, speciation, and radiation. Whereas nonhuman animals come to occupy new ecological niches, humans can create new social niches—new status systems to expand the routes to status. But because social and cultural forces traditionally impose hierarchy, an oppositional force was needed to create new social niches. Enter cool. This is why the first phase of cool, rebel cool, was oppositional.

The rise of cool consumerism was a rebellion against three strands of the Status Dilemma: hierarchical social structure, the psychological motive of status emulation, and the conception of status as having one dimension (specifically, wealth). Beginning in the 1950s, rapidly rising standards of living and growing mass media increased competition for status in a hierarchical society. Social pressures to conform, racial and gender discrimination, and social institutions designed to maintain the status quo all conspired to intensify the Status Dilemma. The emergence of cool stemmed from an oppositional stance that rejected this dominant hierarchical social structure instead of emulating those on its upper rungs. Indeed, the architects of rebel cool, such as Norman Mailer and Jack Kerouac, inverted the dominant social hierarchy, rejecting the values of those at its top and appropriating the values of those at its bottom. The defining quality of cool, like much of modernism itself, depends exactly on rejecting emulation—on seeking to oppose the norms of traditional status—personified by the image of Kerouac ditching Columbia University to head out on the road with the petty criminal Neal Cassady.

As Thomas Frank chronicles in *The Conquest of Cool*, the anti–status quo values of cool aligned with consumption seamlessly and rapidly. Imagine a teenager in the late 1950s donning a leather jacket like the one Marlon Brando wore in *The Wild One*. Wearing it was an act of protest, evoking scorn and contempt from the arbiters of mainstream taste, which would give its wearer a perverse sort of pride. Disapproval from the status quo establishment led to an increase in self-esteem among rebels and respect from the rebels' in-group. This is a form of negative consumption directed at out-groups that's linked to the brain's avoidance system (more on that below). These dynamics helped transform hierarchical social structure into an increasingly pluralist one as more and more diverse lifestyles proliferated. This transition helped reduce direct competition for status by expanding the routes to status, and so alleviated the Status Dilemma. We suspect that the alleviation of the Status Dilemma and the increase in routes to status is one reason why happiness has increased around the world over the last three decades or so.[16]

As these changes reshaped consumer culture, rebel cool itself morphed into a second phase that now reflects the realities of what pundits refer to as our "knowledge economy."[17] For these reasons, the historical period from the rise of cool as an oppositional norm in the 1950s to the present-day DotCool, involving the transition from an industrial to a knowledge economy, has been an age of staggering social rearrangements and societal fragmentation.

To get a glimpse into the rise of cool consumption as a rebellion against the Status Dilemma of 1950s hierarchical society, consider the rise of teenage rock-and-roll bands in the late 1950s, a good illustration of the emergence of cool consumption. As James Coleman recounts in his classic 1961 work *The Adolescent Society*, athletics was almost the only route to status for high school boys of that era. Having such a limited route to status, perpetuated by the social organization of schools, created a Status Dilemma as schools grew larger and competition for limited status intensified. This competition created pressure to diversify social niches. As a result, teens created new routes to status by participating in rock-and-roll bands, as the sociologist William Bielby chronicles in detail.[18]

Traditional accounts of the rise of teen bands emphasize rebellion against parental authority. But the real impetus behind the rise of teen bands was the Status Dilemma that schools created. Indeed, as Bielby notes, Pat Boone, not Elvis Presley, was the favorite recording artist among high school boys and girls of the late 1950s— hardly the figure of rebellion. As Bielby states, "demonstrating competence in rock and roll performance was seen as a potential means of gaining the same kind of peer acceptance as one does from being athletically competent—and again, and in my interviews, it is typically articulated in just that way."[19] While the early rock-and-roll persona may have opposed the norms associated with high school athletics, the culture of rock would become increasingly oppositional later on when it merged with the counterculture movements of the 1960s.

The rise of mass media made it easy for boys joining rock-and-roll bands to present themselves in terms of a rock-and-roll *identity*. The growth of television in particular meant millions of Americans now watched the same images of youth lifestyles. Meanwhile, Hollywood was portraying teenage music lifestyles in movies, such as Alan Freed's *Rock Around the Clock* and *Rock, Rock, Rock* (both 1956), and was revealing a growing fascination with the newly emerging "teenager," as seen in movies like *Gidget*. As Bielby notes, by watching television in the late 1950s a teen could readily learn that a rock-and-roll identity meant wearing white bucks, rolled-up jeans, sleeves rolled up two turns on a short-sleeved shirt, and a skinny belt buckled on the side. Television amplified the meaning of goods and created mass understandings of lifestyles, and people could now easily recognize the teen's adopted rock-and-roll persona.

Mass media weren't the only factors behind the popularity of the teenage rock-and-roll band and its route to status. Postwar economic expansion, growing material affluence, and the rise of middle-class consumer spending contributed critically to the rise of cool consumption. This was an era of dramatically rising standards of living: Average family income increased by almost 60 percent between 1950 and 1960. Unemployment rates hovered at historic lows: the average unemployment rate in the 1950s was 4.5 percent. The upshot was that more and more money was going to feeding lifestyles rather than stomachs. Indeed, consider that in 1900 families spent about 80 percent of their income on necessities (defined as food, housing, and clothing). In 1950, it was about 70 percent, but then it declined rapidly to only 50 percent by 1980 (around today's level). In 1900, U.S. households spent almost half their income on food. A century later, it had declined to about 13 percent. These are all indicators of the century's rise of "lifestyle" discretionary spending.[20] In the 1950s, discretionary teen spending would have a powerful effect on shaping youth lifestyles by merging music and mass consumption, beginning with the rock-and-roll teenager. In the 1960s, though hippies avowedly rejected consumerism, their lifestyle also depended on rising standards of living and the discretionary spending that facilitated music, travel, drug experimentation—even the ubiquitous hippie VW bus. "Alternative lifestyles" were made possible by the rising material affluence of postwar America.[21]

It's no accident that musical genres also diversified over time as consumerism provided more opportunities to build lifestyles around various musical styles. Music and youth lifestyle consumerism are now completely intertwined. Today, young people believe that their musical taste is the best indicator of their identity, an identity that extends to how they represent themselves on social media.[22] Music lifestyles are so clearly demarcated among youth today that there's a deep consensus among young people about what various kinds of music reveal about fans' personalities, values, ethnicities, and even social class.[23] To give you a sense of the diversity of music today, consider the following: Despite predictions of the death of the hit, a relatively small number of songs account for most of music sales today. In fact, the top 1 percent bestselling hits each year account for about 75 percent of all artist revenues. That sounds as if most people are listening to the same songs. But we need to put that 1 percent more into context. That's 1 percent of the roughly 25 million songs for sale. So the top 1 percent is 250,000 songs! To put that in context, you could listen to a different top 1 percent "hit" continuously every day for a year and only get through about half of them.[24] The capacity to construct lifestyle identities around musical tastes helps explain the explosion and diversification of popular music styles, from a relatively homogeneous 1950s rock and roll to a wildly proliferating set of genres and subgenres. There are now more than one thousand distinct musical genres.[25]

These forces have profoundly changed the structure of status systems over the last three decades, particularly among the Millennial Generation (those born between 1982 and 2003).[26] Consider, for example, the modern high school. To many of us, there are few institutions more hierarchical than the American high school. Think of how the high school is portrayed in *Clueless*, for example. An adaptation of Jane Austen's *Emma*, the 1995 movie focuses on a high school where the social hierarchy is as rigid as in Austen's England. The archetypal high school hierarchy is well-known: from the jocks, cheerleaders, and preps at the summit to the druggies and wallflowers at the bottom.

But as pervasive as this image of high school hierarchy is, most contemporary high schools have *much* more complex status relations and are typically more pluralistic than hierarchical.[27] As we've seen with the limited routes to status in the 1950s high school, the traditional hierarchical high school creates a Status Dilemma. Another structural pressure is the growing population of the typical student body today. Ever larger high schools create "structural pressures toward cultural differentiation and pluralism."[28] Such pressures sparked more-pluralist school cultures, in which there is less ranking, stigmatizing, or hostility among groups. In fact, in such school cultures, students frequently mix among members of different groups and move back and forth across them. New routes to status facilitate this proliferation of lifestyles, including the expansion of extracurricular activities such as band, orchestra, chorus, and drama. The popularity of the television show *Glee* (which also emphasizes pluralistic norms) and its cultural impact is a notable example. According to a National Association for Music Education poll of choral teachers, record numbers of students are turning out for auditions as this route to status becomes increasingly visible.[29]

As the sociologist Murray Milner notes, an important element of the increasingly pluralistic structure of high schools is the growing acceptance of gender equality and diversity. In a traditional school's hierarchical setting, a young woman's status was often dependent on her associations with men. In a pluralistic structure, the routes for women to gain status expand to include, for example, being a leader in school government, participating in a broader range of athletic programs, and more. A pluralistic structure also offers increasing access to status and respect for LGBT students.

Gender equality and gender diversity are themes that recur again and again in the transition from hierarchical to pluralistic status. Diane Martin, along with Schouten and McAlexander, for example, examined the changing nature of women's participation in Harley-Davidson groups, noting the pronounced move from the backseat to the driver's seat and that move's redefining role in those microcultures.[30] So where do our concerns about status come from?

Status Systems in the Wild

In 1651, Thomas Hobbes asked, in his work *Leviathan*, what life might have been like in a state of nature prior to the advent of government and political authority. He imagined it would have been a war of all against all—for two essential reasons. First, he argued, all mankind endlessly seeks glory, or status. Even a powerful king is never satisfied with his empire. Second, people are equal in terms of their capacity to kill one another, giving all equal claim to resources. The result is a life that would be, as he famously put it, "nasty, brutish, and short." There is considerable controversy regarding what Hobbes meant by the state of nature, since he used the term to refer interchangeably to a hypothetical pre-civilization era and to the breakdown of contemporary society from civil society to anarchy, a condition he was all too familiar with from the English Civil War. Hobbes's speculations about the "savages" of America, however, suggest that he intended the term to include a time before political authority. Today, speculations about the state of nature extend not only to human evolution and prehistory, but also to the evolutionary links between human and nonhuman social structures.

As it turns out, Hobbes's speculation that there is a natural equality among humans—namely, an equal capacity to dominate others—doesn't hold true for our nearest genetic relative, the chimpanzee. And therein lies nature's solution to the Hobbesian state of nature—a solution that leads to a different outcome than Hobbes imagined. Chimpanzees compete intensely with other chimpanzees for resources, particularly food and mates. If chimpanzees were essentially equal in terms of their abilities to acquire food and mates, then their life likely *would* be a Hobbesian war of all against all. But male chimpanzees aren't equal in Hobbes's sense. They differ just enough in strength and other advantages to make them unequal in their capacity to claim valued resources.

Based on these differences, male chimpanzees sort themselves into a linear social hierarchy so that, roughly, a member of the group that is higher in the hierarchy is dominant over those below it. The essential feature of their social hierarchy is that disputes over resources are often avoided by subordinates, who either flee when a dominant male challenges them, or exhibit a submissive signal. In an ironic, Hobbesian fashion, the alpha male also acts as a kind of Leviathan, who maintains order by quelling disputes among community members.[31] The status system in the wild establishes a pecking order for access to resources, where some members have more claim to those resources than others. Hierarchy is nature's solution to the problem of social order, and this is likely the primordial source of social inequity in human societies.

The Rebel and the Noble Savage

Social status hierarchies are ubiquitous across species ranging from insects to birds to fish and mammals. But what about us? Many consumer theories still hold on to the idea that status concerns are modern society's "unnatural" creations. In so doing, they often assume some version of Rousseau's Noble Savage. Today, this takes the form of speculating about existing premodern societies, hunter-gatherers, and their near-universal egalitarianism. That is, present hunter-gatherer bands typically have weak political leadership, and rank and stratification are typically absent. For these reasons, consumerism's critics often romanticize hunter-gatherers as pacifists, having few wants and seldom quarreling because they're free of status concerns.[32]

But these views misrepresent the nature of hunter-gatherer egalitarianism. The Noble Savage view is also simply wrong in claiming that social stratification was never present before the emergence of complex societies such as chiefdoms, primitive kingdoms, or early city-states. Despite the prevalence of egalitarianism among hunter-gatherers today, social stratification existed in societies far earlier than the advent of agriculture and the transition to early city-states. For example, grave goods at the archaeological site of Saint-Germain-de-la-Rivière, about thirty kilometers east of Bordeaux, dating to around 15,500 years ago, reveal exotic personal ornaments that likely marked privileged social groups.[33] Indeed, archaeological evidence reveals that non-egalitarian hunter-gatherers have existed across the globe.[34] So does contemporary hunter-gatherer egalitarianism reflect a basic lack of status concerns, an innate egalitarianism like that which some other social primates, such as squirrel monkeys, appear to display?[35] Or are we like chimpanzees, chickens, and many other animals that are innately despotic, whose nature drives them to compete for position so that their status rivalry results in social hierarchy?

The anthropologist Christopher Boehm has long been interested in these questions as they relate to primates and hunter-gatherers.[36] His research provides important insights into how hunter-gatherer bands maintain their egalitarian structure. Boehm argues that humans are innately despotic and not naturally egalitarian. Hunter-gatherers maintain egalitarianism by what he describes as a reverse dominance hierarchy and leveling mechanisms. Coalitions of subordinates intentionally curtail alpha-type behaviors through social sanctioning of upstart power seekers, a strategy Boehm calls "subordinate rebellion."[37] Boehm notes that an essential feature of a despotic species is a strong dislike of being dominated, which makes it tempting to rebel against superiors. Thus, subordinates can form coalitions that collectivize and institutionalize their rebellion against individuals who seek to dominate others.

The dislike of being dominated is the primordial source of the rebel instinct to oppose dominant hierarchies. Boehm suggests that this tendency to rebel is related to the coalitions chimpanzees sometimes form against alpha males, and notes that the same tendency was likely present in the common ancestor we share with them.[38] This coalition behavior—primordial collective action—became more elaborate in humans and ultimately became intentional. He argues that this may in fact have been the origin of human morality, as in human societies moral sanctioning is the predominant method of keeping alpha male behavior in check, directed at those males who create undue competition, domination, and victimization. Often the result is major tension between the group and the status-seeking individual.

It's not clear how the rebel instinct is activated in the brain, but we suspect it's related to the strong emotional reactions we have to many forms of unfair treatment. For example, recall the ultimatum game we introduced you to in chapter 3 involving whether or not to accept a person's offered split of $20. Now, suppose he offers you $2 (you know he was given $20). Would you agree to the split? Most people find such a low offer insulting and would rather not get anything. Many economists regard the rejection of unfair offers as an economic mystery, since accepting even a low offer maximizes utility—it's better than nothing.[39] One explanation for why people reject these low offers is that it's an evolved response that may be costly, but enforces cooperation and helps people avoid becoming the sucker others can take advantage of. For these reasons, it's likely the product of the Survival pleasure machine. Think of it as a social instinct that's an emotional response to an evolutionarily important problem. That's why people reject the offer without having to think about it. Looking inside the brains of people who were given lowball offers, researchers found that activity in the insula related to how likely people were to reject an offer.[40] The greater the activity, the more likely it was that they were going to reject it. This area is involved in emotions such as disgust, and scenarios that involve social revulsion often activate it (it's probably no accident that the language we use for rebellion against injustice overlaps with the language we use to describe the viscerally disgusting). In contrast, when people are told the offer comes from a computer program that decides the amount to offer randomly, people are quite happy to accept low offers and don't get mad. This reveals that it's not the offer per se, but the bad intent behind it that angers people. It's the sense that someone is trying to take advantage of you that leads to spite, grudges, and feuds—all emotionally laden Survival pleasure machine behaviors we pursue despite their not only not benefiting us but often costing us greatly.

Evidence suggests that children are particularly attuned to unfairness, as parents who anguish over the equality of holiday presents for their children know only too well. Indeed, some nonhuman primates also resent being treated unfairly, and will even angrily toss a bit of food back at the person offering it if that person just offered a larger piece to a nearby primate.[41] While we've referred to this as an instinct, behavioral dispositions interact with social contexts as well. These social contexts can have dramatic effects on our behavior, giving us the flexibility to adapt quickly to environmental changes. When defensible resources exist, innate despotic tendencies often result in a rapid transition to social hierarchy. The instinct to rebel may be tempered by the potential gain from centralized political structures. Social psychologists have studied this taste for hierarchy under the rubric of social dominance orientation—that is, individual preference for group hierarchy and inequality. They measure this trait by asking people to state their acceptance of such statements as "Some groups of people are just more worthy than others," or "It's probably a good thing that certain groups are at the top and other groups are at the bottom."[42]

We share the basic neural mechanisms of status with many other social animals. Status changes and status anxiety have effects on virtually every system of the human body, effects as severe as cognitive impairment, suppressed immune function, hypertension, elevated levels of stress hormones, and decreased fertility in both sexes.[43] Brain signals in response to status threats flow to glands of the endocrine system, including the pancreas, thyroid, pituitary gland, adrenal glands, ovaries, and testes, which in turn release hormones that regulate growth, metabolism, reproduction, and responses to stress and injury. Social status conflicts and the effects of psychological stress have far-reaching implications for human health. Indeed, the social epidemiologist Michael Marmot, studying what he describes as the Status Syndrome, found that the lower in a hierarchy a person is ranked, the worse their health, even after accounting for differences such as access to health care and other risk factors. He argues that status differences underlie chronic stress from lack of autonomy and social participation.[44]

Another indication that our biology reflects the centrality of status to our lives is the fact that we can recognize a person's social status from facial cues, such as a slightly lifted chin, in a fifth of a second.[45] We also rapidly infer social status from postures that appear to be similar across cultures, such as physical expansion (dominance) or contraction of the body (submission). We share these abilities with other primates, for whom quick recognition of status is likewise essential for navigating socially stratified environments.

The biological reality of our status concerns came into striking focus for us a few years ago when Steve's lab collaborated on a study with Read Montague's lab at Baylor College of Medicine.[46] The idea was to examine how a person's perceived status affects their learning. This topic has many real-world connections. For example, many elementary school teachers group students on the basis of reading or math ability.[47] These groups create potential Status Dilemmas, since your status in the class depends on which group you belong to, and kids are good at figuring out how the groups differ. So the experiment had a small group of people take an electronic IQ test together on individual screens while their brains were scanned. After each question, each member of the group saw his or her relative ranking in the group, based on everyone's answers. As you might imagine, people found this pretty stressful. Before each participant saw their ranking on the screen, there was strong activation in their amygdala (which relates to fear). But what was most interesting was what happened in a person's brain when their rank went up: the nucleus accumbens (the region of the brain that responds to reward) activated strongly, as it does when someone receives some cash or a hit of cocaine.

The fact that people didn't receive money or a prize for doing well in the test raises a fundamentally important point. Some economists and consumer critics assume that our status concerns aren't intrinsic—that is, that we don't actually care about status per se. We only care about it when it's a means to things we do care about, such as coveting an expensive house for the advantages its school district affords one's children. The activation of the nucleus accumbens in this experiment proves otherwise. The group knew they would disband right after the test and the members wouldn't interact in the future. So a high ranking in the social IQ test didn't lead to any tangible benefits. Yet the participants still cared a great deal about their rankings and found increases to them rewarding. Hence, status is intrinsically rewarding.

Status in Transition

An unlikely source recently bolstered the possibility that the evolutionary logic of social selection drives status consumption.[48] The Tsimané of the Bolivian Amazon are a group of indigenous people who live in approximately eighty villages deep in the lowland forests east of the Andes. While their villages are in some of the most remote places on earth and many continue to live in forager-horticulturalist groups, they are increasingly in contact with the outside world. Specifically, they are now experiencing a transition economy from foraging and subsistence farming to cash cropping and wage labor.

Examining Tsimané consumption provides us with insight into an economy that is beginning to feature incomes and consumer goods and some discretionary monetary expenditures, but is not yet immersed in other elements that may drive spending, including marketing and advertising. This provides key information, since most studies of consumption focus on industrial societies. In a study of Tsimané consumer habits, the anthropologist Ricardo Godoy and his colleagues looked at how approximately four hundred Tsimané in thirteen villages spent their money across a number of goods, including clothing, health, transport, kitchen utensils, and luxuries, and whether more money was spent on goods that were visible to others. What they found was that a higher share of expenditures went to what are considered luxury goods among the Tsimané, such as watches, mobile phones, and backpacks. These results suggest that signaling with goods—just as we saw with the fifteen-thousand-year-old shell necklaces—emerges whenever conditions allow it and *precedes* institutional forces such as marketing and advertising, which we often think of as *creating* concerns about status.[49]

From Status to State

Evolutionary biologists argue that animal social organization, and particularly that of primates, is rarely fixed. This idea fits very well with our cultural biology perspective. Primates are remarkably flexible, altering their social organization in response to ecological conditions. Female chimpanzees, unlike female bonobos, seldom form coalitions in the wild, and are subordinate to even low-ranking males. In contrast, in captive chimpanzee groups, females form strong bonds that result in coalitions that allow them to sometimes dominate high-ranking males. In humans, it appears that the economic defensibility of resources, whether they are marine resources such as salmon runs, arable lands, trade routes, or livestock herds, is a key condition for the emergence of stratified social organization.[50] For one thing, the forms of wealth that many hunter-gatherers possess (embodied, as in physical assets, and relational, as in their social connections) permit of only a low degree of intergenerational transmission, so accumulated inequalities are limited across generations. Defensible resources, in contrast, can be passed from one generation to the next via inheritance.

Whenever defensible resources exist, it seems humans have abandoned egalitarian norms in favor of ones that allow resources to be inherited, creating institutionalized inequality and hierarchy.[51] These hierarchies typically emerged rapidly because of our innate dispositions for status, alongside a desire for political authority to preserve order and protect accumulated property.[52] Expanding systems of social stratification further systematized social classes and regulated movement among them. The prevalence of extraordinarily complex sumptuary laws is a good example of institutionalized efforts to enforce social hierarchy.

Sumptuary Laws

For many societies throughout human history, the meaning of goods was fixed not only by local traditions but by sumptuary laws. Status itself was typically ascribed rather than achieved, whether by a formal caste system or merely by the limited roles available to one's status as defined by class, gender, or race. An individual's consumption was not just limited by available resources, but by the pattern of consumption that was allowed under law according to ascribed status. One result was that prior to the rise of a consumer economy, socially visible goods transparently relayed their meaning with little room for ambiguity.

Sumptuary laws were aimed at regulating outward signals of dress, either as formal laws or as norms governed by religious codes. Although we may be tempted today to dismiss clothing as frivolous fashion, the richness of clothing and personal adornment are signals of economic status as well as of social and deep moral norms, from the neckline to the hemline. Consider how much anxiety "casual Fridays" have caused by confusing the norms surrounding appropriate office apparel. As the fashion historian Aileen Ribeiro explains, there is a deep and continuing connection between dress and morality that was historically regulated by the church or the state.[53]

The social categories that sumptuary laws enforced included economic ones, but also moral ones, ranging from the attire an upstanding man or woman would wear to categories of moral deviancy, including clothing reserved for the prostitute, the convict, the noncitizen, and the nonbeliever. The earliest written Greek legal code, the Locrian code in the seventh century B.C., included laws linking signals and status: embroidered robes, for example, were only to be worn by prostitutes. The toga was highly regulated under Roman sumptuary laws to signal one's social rank and age by its color and the width and number of stripes along its border. Feudal Japan was said to have the most intricate set of sumptuary laws, which tightly regulated clothing according to social class. In the eighteenth and nineteenth centuries, when the merchant chōnin class grew wealthier than the aristocratic samurai, these sumptuary laws kept in place social distinctions despite the economic ability of the chōnin to acquire aristocratic goods.

In fourteenth-century England, laws were introduced with the explicit goal of curtailing people from dressing above their station. For example, farmers could only wear coarse woolen fabrics; men who were not knights but received rents in excess of £200 were allowed to wear coarse silks and some furs, whereas those earning less were restricted to cloth. Knights could wear most silks except cloth of gold, and most furs except ermine, which were both reserved for royalty.[54] A simple glance at someone was enough to know their social standing and occupation. Both the Islamic world and medieval and Renaissance Europe had complex sumptuary laws, including requirements of minority religious groups to wear badges, hats, bells, or other items of clothing that distinguished them from members of the dominant religious group—some of which were revived by the Nazis, such as the yellow patch. Under the Taliban in Afghanistan, members of the Hindu minority were likewise required to wear yellow patches.

These laws were almost universally resisted. Wherever economic conditions allowed, people wanted access to goods that were only available to those of higher station. In Japan, the merchant chōnin class demanded revisions to the sumptuary laws and were eventually granted concessions. In other areas, such as sixteenth-century England, these laws were often ignored and widely broken. It is difficult to reconcile the fact that historically people routinely violated or opposed sumptuary laws with the sociological view that consumption is driven by advertising and marketing manipulations of socially induced wants. In the absence of marketing and even in the presence of laws against it, people throughout history have sought out status goods whenever they had the means to do so.

Status Wars?

Viewing our status and rebel instincts from an evolutionary perspective is unsettling, if not downright depressing. Some economists, notably Robert Frank, argue that humans are so consumed by status seeking that we live in what he's dubbed a Darwinian economy, and the only escape is a steep progressive tax on consumption.[55] Let's look a bit deeper into the lives of our closest genetic relative to see how the status instinct and the rebel instinct play out. We suspect that the human parallels will be obvious—and ominous.

Consider an event that took place on the morning of October 2, 2011, in the mountains of Tanzania.[56] Pimu was a twenty-three-year-old chimpanzee who had been the alpha male for four years. The male dominance hierarchy was stable, with low rates of male aggression. On that morning, Pimu got into a brief but fierce fight lasting about half a minute with the second-ranking male, who bit Pimu badly on the face before running away. Four adult males who had observed the fight suddenly charged at the injured Pimu. The four males surrounded him and began beating and biting him. One male broke off a large branch and used it to pin Pimu down while others bit his arms and legs. Other chimpanzees came in support of Pimu, but were chased off. For forty-five minutes, the attack continued, punctuated by brief bursts of attacks followed by lulls during which the attackers would let out loud, agitated hoots. Pimu, unable to escape and badly injured, pant-grunted to one of the attackers, a sign of surrender. Instead, the four males attacked him one final time and then withdrew into the surrounding trees. Lying motionless, Pimu died a few minutes later. One of the instigators of the attack was the third-ranking male, leading to speculation that the attack was an opportunistic one. His coalition saw a moment of weakness and exploited it.

Lethal attacks like these are rare among chimpanzees. But this rarity is a dark reminder of the power of the status instinct. Male chimpanzees are typically a noisy lot, but sometimes a group of them will turn silent and leave the group. They form a single-file line and head out quietly toward the borders of their territory. There, they patrol with purpose, looking for chimpanzees from nearby communities. If they come upon a male who has strayed a bit too far on his own, they will attack him. These attacks are ferocious and lethal. Researchers are confident that these are turf wars. Chimpanzee groups battle one another for territory and its food, sometimes even taking over a group's entire territory and exiling them. The fact that male chimpanzees rely on each other to defend and expand their territory is one reason why killing within the group is rare. Whatever gentle images of our chimpanzee cousins we grew up with, the reality is that their behaviors reflect how profoundly status seeking dominates their lives, from their struggles for power within the group to warfare over power with neighboring groups.

The extraordinary preoccupation with status among male chimpanzees begs for explanation. Are we like our closest genetic cousins—trapped in a status arms race?

Rich Ape, Poor Ape

Chimpanzees compete intensely for status within their group and cooperate to compete against other groups for status. Why are they so preoccupied with status? There seem to be two driving factors.[57] First, female chimpanzees have extremely low reproductive rates, giving birth once every five to six years. This means the number of sexually mature males dwarfs the number of reproductively active females. Female chimpanzees are also sexually receptive only for about two or three days every cycle, further limiting mating opportunities. Alpha males sire about 30–50 percent of all infants, so there are large fitness benefits to being the top ape. Higher-status members get better food patches, which are in short supply, as evidenced by the intergroup conflicts over territory for food. Female chimpanzees also appear to compete with one another for food, and so do not form coalitions, with the exception of some female chimpanzees joining together to kill the babies of immigrant females to reduce competition for food (female chimpanzees leave their birth community). No wonder male chimpanzees are preoccupied with status. It's also important to point out that being the alpha male involves costs. In particular, alpha males spend a great deal of energy guarding reproductively active females from would-be suitors and supporting coalition partners. For this reason, they experience a lot of stress, as revealed by their high levels of the stress hormone cortisol.[58]

Compare the chimpanzee to the bonobo, which diverged from chimpanzees about 1 million years ago.[59] Although it's not a moniker we completely agree with, bonobos are sometimes referred to as "hippie chimps" because they seem to embrace a "make love, not war" ethos. They are less aggressive than chimpanzees within the group and do not express status through the sorts of signals we saw with chimpanzees. While bonobos have social hierarchies, they differ in important respects from chimpanzees. Most notably, female bonobos are slightly dominant or co-dominant with males, and female alliances are strong. As a result, males are not typically aggressive toward females, as is the case with chimpanzees. In fact, male and female bonobos form bonds. Males have a hierarchy among themselves, but a male's rank is typically dependent on his mother's rank. Perhaps the most intriguing difference between chimpanzees and bonobos is their sexual behavior. Bonobos engage in elaborate "sociosexual" behavior, meaning that sex plays a social role in addition to a reproductive one. Genito-genital rubbing between female bonobos is common, as is male-male mounting and even male-male oral sex. Females are sexually receptive for most of their cycle, which suggests concealed ovulation like that of humans. Male bonobos do not mate-guard females or block matings by other males, and their contacts with outside groups are also more peaceful than chimpanzees' warlike tendencies.

What explains the extremely different lives of bonobos and chimpanzees? The Harvard primatologist Richard Wrangham has a provocative theory.[60] Bonobos live south of the Congo River, while chimpanzees live to its north. Chimpanzees thus share their habitat with gorillas, who are voracious plant eaters, eating as much as forty pounds of leaves, stems, roots, vines, herbs, trees, and grass every day. As a result, chimpanzees rely more on fruits, with about half their diet consisting of figs alone.[61] The difference means that chimpanzees spend most of their day in so-called foraging parties that differ markedly from those of bonobos. Chimpanzee communities divide into relatively small parties to forage (each party contains 9–30 percent of the community), whereas bonobos divide into larger foraging parties (25–50 percent of the community). Unlike female chimpanzees, who often forage alone, bonobo females forage together and in mixed-sex groups. As a result, female bonobos form close alliances, which over evolutionary time deeply altered their status relations.

Bonobos, then, are richer than chimpanzees in terms of the two currencies that matter most to them: food and sex. The bonobos' absolute wealth decreases the benefits of relative position. A bonobo community isn't a rankless commune by any means. There are still status hierarchies, but the "return on status" isn't nearly as limited to the upper rungs as it is for chimpanzees. It matters much more for a chimpanzee to be at or near the top of its status hierarchy.

It makes for interesting speculation to ponder whether we are more like chimpanzees than bonobos. On the one hand, there's evidence that humans have "self-domesticated" like bonobos, meaning we retain juvenile characteristics longer.[62] Human females also share bonobos' concealed ovulation. On the other hand, the legacy of intergroup competition, including our in-group favoritism, points to commonalities with chimpanzees.[63] The most likely possibility, however, is that our increased behavioral malleability means we alter status relations flexibly through norms and cultural institutions. One notable example of this is the cultural development of monogamous marriage.[64] In about 85 percent of all known human societies, men were allowed to have more than one wife. The rise of hierarchical societies typically coincided with increased levels of polygynous marriages (multiple wives) as more wealth accrued among the elite. The global spread of monogamy is fairly recent—Japan prohibited polygyny in 1880, China in 1953, and India in 1963. Polygyny intensifies status competition and increases the number of unmarried men—it creates a Status Dilemma. As the anthropologist Joseph Henrich and his colleagues suggest, monogamous marriage was a cultural innovation aimed at alleviating status competition. It had far-reaching social consequences by reducing rape, murder, assault, robbery, fraud, and domestic abuse. Monogamous marriage also results in lower birthrates and less gender inequality while increasing parental investment, savings, and economic productivity. It also has a direct effect on our biology in terms of reducing testosterone, effectively dialing down the status instinct.[65] What's curious about the advent of monogamous marriage is that it would have been against the interest of the ruling elite, since they had the most to lose. As Henrich and his colleagues suggest, the pressures of intergroup competition may have been so great that these interests were traded off for the sake of social cooperation.

Monogamous marriage is one human innovation that defuses human status competition and so alleviates the Status Dilemma. As we'll argue in the next chapter, oppositional cool consumerism is another.

6

DARWIN GOES SHOPPING

While some people recoil at the idea of applying Darwinian principles to our economic life, we think the problem is that economic theory isn't Darwinian enough. Let us explain. The image of our economic life as a Darwinian competition for limited resources is only half the Darwinian picture: his principle of selection. To appreciate the other half, let's look at a curious experiment that inspired Darwin's other great principle. It took place in the gardens of Woburn Abbey, north of London. In the 1820s, George Sinclair, the abbey's head gardener, planted grasses on two equal-size plots of land. On one, he planted two varieties of grasses. On the other, he planted twenty varieties. If one grass's success would be another grass's failure—locked in a competition for sunlight and nutrients—we'd expect the intense competition among the twenty species of grasses to result in a diminished harvest. But what Sinclair discovered was remarkable. The plot with twenty species of grasses produced almost twice the harvest of the plot with two grasses (a crucial insight today for the importance of biodiversity).[1]

This result was key to what Darwin would call his principle of divergence. The extraordinary bounty of life in the plot with twenty grasses stemmed from the fact that each species required different resources for growth, such as various nutrients in the soil. As a result of diverging from one another, they diffused their competition.

So while natural selection eliminated the unfit, divergence drove species away from competing with one another—to become different and diverse. It was, Darwin surmised, nature's version of the division of labor.

Indeed, Darwin witnessed divergence in action when he visited the Galapagos Islands. In particular, he kept encountering birds that looked similar but had all sorts of beak shapes that allowed for different diets. He returned to London and had the leading ornithologist of his day, John Gould, examine them. Gould pronounced the birds to be various finch species—a dozen in all—and they were unlike others anywhere else in the world. Darwin would go on to theorize that a single species likely came to the islands from the mainland. The islands, Darwin continued, featured new niches, containing different types of nuts, that hadn't been available to them on the mainland because other species already filled those niches. In other words, the islands were rich with ecological opportunity.[2] Rather than compete for the same resources, the species diversified—not purposefully, but through mutation. Darwin would call this process of divergence adaptive radiation.

Evolution's exuberance makes it impossible to know today how many species exist. It's unknown whether there are a few million species, 5 million, or perhaps 10 million or 20 million, due to the astounding abundance and ubiquity of living beings, spread across nearly every nook and cranny on earth. There may be more than a million different species of beetles alone! Whatever the number, we do know that it is only a tiny fraction of all the species that have ever lived, even long before anything human came around.

Ecological opportunity can occur through other means as well. For example, the number of different types of horses reached its peak of about a dozen in North America approximately 10 million years ago after climate change had expanded the grasslands, opening up new niches for grazers. Another example of ecological opportunity stems from the extinction event some 65 million years ago, known to scientists as the Cretaceous-Paleogene, or K-Pg, extinction, when a six-mile-wide asteroid striking the Yucatán Peninsula (or, in an alternative scenario, catastrophic volcanic eruptions) caused the extinction of approximately 75 percent of all species, including the great dinosaurs. This mass extinction was followed by many radiations as surviving species moved into and adapted to niches formerly occupied by species now extinct. Key innovations among animals, such as wings, also provide ecological opportunities for species to take advantage of new resources. While adaptive radiation typically occurs by colonizing available niches, sometimes organisms will modify the environment themselves, creating new niches that their own species or others will take advantage of. We can observe experimental adaptive radiation in the laboratory by growing bacteria that create new niches, which are occupied by new bacterial species.[3] Altering the environment in this way is known as niche construction—nests and dams are two examples. Human evolution has no doubt been influenced by niche construction. In our case, however, many of the niches are cultural and social.

We don't have to go very far to see how divergence affects us. If you've ever wondered why your personality differs so much from those of your siblings or why your kids are so unlike one another—despite some degree of physical resemblance—one answer is that competitive pressures operate in a household that are similar to those in any ecosystem.[4] In fact, your personality is about as similar to a sibling's as it is to that of a random person walking down the street.[5] As we saw in chapter 4 in the case of the spotted hyena cubs competing for mother's milk, sibling competition for parental affection is a pervasive evolutionary force. It remains so today. The parenting sections of bookstores are filled with books dealing with sibling rivalry, while the self-help sections are filled with books on the emotional scars left by parental favoritism!

Children can compete for parental affection and approval in head-to-head contests, but that's a tough gambit. Just ask the NFL quarterback Eli Manning. The younger brother of the NFL quarterback Peyton Manning, Eli has endured years of unfavorable comparisons to his older brother despite a brilliant career. He's routinely asked if he's jealous of his older brother and even who his parents like more.[6] Siblings instead tend to make themselves stand out from one another to diffuse direct competition, both by developing distinct personalities and by occupying different family niches. Spacing between siblings also has an effect, with siblings who are closer in age becoming more different due to increased competition. For example, only two years separate Peyton and his older brother, Cooper. Cooper was a wide receiver in high school, not a quarterback. The reason was that despite being a little younger, Peyton was already the better quarterback, was on the same high school team, and in fact threw Cooper most of his passes. Cooper diversified. Eli, five years younger than Peyton, never overlapped with him on a varsity team, and so the quarterback niche was open. The fact that their father was also an NFL quarterback and remained involved in football after retirement also suggests that being a quarterback was a great way to receive parental affection. Many parents try to defuse sibling rivalry by taking the opposite tactic. They promote ecological opportunity by encouraging each of their kids to focus on what that child is good at.

Being the firstborn also means you're more likely to be the dominant child as well as the "mini-parent" to your siblings. You're the enforcer. Later-borns are more likely to break the rules, are more liberal, and take on more risk to find their own niche. As a result, they tend to play more dangerous sports. Among brothers in professional baseball, younger brothers are ten times more likely to steal bases. In studies of political and scientific revolutions, Frank Sulloway found that later-borns were more rebellious, often leaders of revolutions, while firstborns were among those most opposed to radical changes.[7] In the competition for parental affection and investment, sibling divergence increases total affection. Direct comparisons become harder to make, and sibling rivalry is reduced, so there's less envy. In other words, if the family were a garden, sibling diversity would make everyone taller.

We believe that the kinds of processes that drive sibling diversity also help drive the proliferation of lifestyles and consumer microcultures. That is, consumerism's adaptive ingenuity is the way it diffuses competition for social status. We intrinsically care about status. When the routes to status are limited, we will compete directly for it or be limited in our access to those routes by subordinating forces, as in a traditional social hierarchy ruled by elites. When the routes to status are open, or when we can create new ones, we will often choose to pursue these routes to avoid direct competition. Doing so diversifies status and increases its total amount in society. A consumer lifestyle or a consumer microculture is its own social niche. Each niche has its own values and norms about status. Being a part of a social niche creates both individual self-esteem (status) and collective self-esteem (respect). Think of status pluralism as a social version of Sinclair's garden.

As we reflected on the new findings that rising purchasing power increased happiness, we grappled with whether that happiness increment had any connection to lifestyle diversification. At the time, Steve was teaching a class on distributive justice, which deals with how society should distribute its benefits and burdens (with Ming Hsu and Cédric Anen, he'd published a brain-imaging study on the topic a few years before).[8] The topics seemed light-years apart, but a remark by the influential American philosopher Robert Nozick, a major contributor to work in distributive justice, suggested an extremely provocative connection. Nozick was discussing the problem of envy when he wrote, "The most promising ways for a society to avoid widespread differences in self-esteem would be to have no common weighting of dimensions; instead it would have a diversity of different lists of dimensions and of weightings."[9] But how would a society actually go about doing this? The more we thought about it, the more we realized it would look a lot like a diversified consumer society. Consumer microcultures create these dimensions—ways that people and groups can be different from one another. Nozick's reference to "no common weighting" means that there's no consensus over how lifestyles are ranked, which is a feature of lifestyle pluralism. Indeed, Nozick's phrase "a fragmentation of a common social weighting" would be a good way to describe the last three decades.[10] If there's only one dimension, then we'll see a hierarchy develop in which every person is envious of those above them. That's a lot of envy, and it's what makes status *look* like a fixed resource.

Try the following thought experiment: Imagine if the only sporting event was the 100-meter sprint. You'd have an intense sports Status Dilemma (to say nothing of the reduced Olympics revenue from an event that lasts less than ten seconds!). A few people are going to get all the glory and all the sponsorships. Notice that even if everyone in the world doubled in speed, it would be the same hierarchy. Relative time, not absolute time, seems to be all that matters. But suppose we introduce a second event—the mile. Now people who were bad at the sprint have a chance to be good at the mile, since endurance running and sprinting call on different strengths. There may be endless barroom banter about whether the 100 meters or the mile is the better event. But people enjoy endless barroom banter like this precisely because there's no right answer. You never find a barroom argument over facts! Certainly Hicham El Guerrouj, the current record holder in the mile, has no shortage of status and likely has no reason to envy Usain Bolt. We'd bet he's a lot less envious of Usain Bolt than is the second-fastest sprinter in the world, who compares himself to Bolt along the same dimension. As we continue to increase the number of sports, we continue to increase the total amount of status available in the sporting world.

There's good reason to think this is what's happened in sports over the last hundred years or so. Not only does the number of sports increase, but each sport increasingly diversifies. Look at the bewildering number of swimming events, or all the events involving people skiing down the hill backward in the Sochi Olympics. A professional cyclist riding the Tour de France can contend for the yellow jersey (overall lead), the green jersey (fastest sprint finisher), the white jersey (best young rider), or the polka-dot jersey (best climber). Sprinters don't envy the yellow-jersey wearer, as a sprinter knows he doesn't have the right build to win the whole thing. Sprinters envy other sprinters. If you're too big to be a sprinter on the road, you can be a sprinter on the cycling track. England's Chris Hoy was knighted for his exploits as a cycling track star. Or you can be an extreme sports star—skateboarding, BASE jumping, ice climbing, kite surfing, snowboarding, surfing, free diving, or just parachuting from space.

As the number of sports niches increases, athletes start resembling Darwin's finches more and more. It's what sports scientists refer to as the "Big Bang of body types."[11] That is, athletes' bodies become more and more specialized for their event, not through evolution but from sports organizations casting a wider net to recruit athletes from every corner of the globe. The best runners have extremely long legs and short torsos, while the best swimmers have extremely long torsos and short legs. Hicham El Guerrouj is seven inches shorter than Michael Phelps, but they wear the same length pants. Basketball players have extremely long arms for their height. Water polo players have long forearms, which lets them fire the ball. Steve was an amateur cyclist, but didn't go on to win Grand Tours like some of his former teammates because of his relatively short femurs (or so he tells himself). As athletes become increasingly specialized for their sport, they become more unlike athletes in other sports, and so there are fewer common dimensions for comparison.

Divergence and Its Discontents

Although diversifying lifestyles is a potential way to confront the Status Dilemma, there are at least two main forces opposing it. First, adherents of existing lifestyles may resist diversifying lifestyles that clash with their cultural values. Second, elites will resist divergence when resources are scarce. We've already seen that social hierarchies tend to emerge as soon as scarce defensible resources appear. Social elites often devise cultural ideologies and institutions that are "extractive."[12] That is, they tend to amass wealth for the ruling class while subordinating others. According to the political scientists Ronald Inglehart and Wayne Baker, all known preindustrial societies have typically been low in levels of tolerance for diversity, and have restricted rights relating to abortion, divorce, homosexuality, and gender equality. All such societies were authoritarian.[13] There are competing views on whether people came to accept the resulting social order. People rationalize their place in such societies, in part because making peace with the status quo makes their world feel less uncertain and less ambiguous.[14] By using measures of social dominance orientation, the degree to which people oppose equality, we can see whether dominant and subordinate groups accept social hierarchy—does social hierarchy exist by consent, or do subordinates dissent from it? In a large analysis of more than one hundred studies that included participants from South America, North America, southern Africa, the Middle East, central Asia, East Asia, eastern and western Europe, and Britain, researchers found that members of subordinate groups strongly reject hierarchy.[15] This rebellious discontent is particularly acute in societies that advocate norms of equality, suggesting that such norms heighten the resentment and sense of relative deprivation people feel when subordinated.

We've seen that one difference between chimpanzees and bonobos lies in the intensity of their status instinct, which reflects variations in the absolute wealth of their environments. Our greater behavioral flexibility depends in part on our ability to alter the intensity of these social instincts in response to changing environmental circumstances and norms. One of the most robust findings in the behavioral sciences is that changes in circumstances—and in our perception of them—have profound and predictable effects on the intensity of our status instinct. The intensity of within-group competition for resources can be predictably manipulated by altering the supply of a resource and its allocation.[16] We already saw this with monogamous marriage, and how its inclusive norms and institutions reduced status competition. Consider also the unintended consequences of China's one-child policy, which started around 1979. Beginning around 1988, China's crime rate started to rise dramatically. By 2004, it had doubled. China's one-child policy had unintentionally created a glut of millions of "surplus" men due to parents' preference for boys. The short supply of potential wives intensified status competition for marriage and doubled the number of young, unmarried men.[17] The link between low-status unmarried males and violent crimes is a constant theme across societies and time. Indeed, it's confirmed daily in our headlines. We see how a scarce good or a narrow niche intensifies competition for more abstract forms of status as well. In political primaries, for example, candidates within the same party often engage in such fierce competition to win their party's nomination that they seriously damage the party's chances in the general election.

These examples reveal that the intensity of our conflicts and the fervor of our desire to subordinate others depends on our perceptions of the supply of what's being fought over. In other words, the less there is of some valued resource, the more intense the conflict. Psychologists dubbed this "realistic conflict theory," and a classic experiment illustrates the idea.[18] In 1954, in Oklahoma, two separate groups of eleven boys were brought to a Boy Scout camp in Robbers Cave State Park. The boys were about eleven years old and didn't know one another prior to their trip to camp. The two groups were housed in cabins far apart from one another so they wouldn't suspect the other group's existence. The experimenters, headed by the social psychologist Muzafer Sherif, masqueraded as camp staff so they could observe and record the boys' behavior. In the first phase of the experiment, the two groups were kept separate and were encouraged to pursue group goals, such as making group meals, going on group treasure hunts, and so on. Each day, researchers wrote down what they thought the order of the hierarchy was. When they compared notes, they discovered they largely agreed. In short order, both groups sorted themselves into status hierarchies.

Soon, the boys began to suspect they might not be alone. They asked whether there were others, and began to talk of defending their turf from them. Their suspicions were confirmed when the two groups were brought together for a multiday sports competition. The winners would get a trophy and a few small prizes (medals and pocketknives). The losers would get nothing. As the games went on, the two groups, now calling themselves the Eagles and the Rattlers, routinely name-called, threw food at opposite-group members during meals, destroyed the flag from the opposite group, and even raided each other's cabins and stole whatever prizes they could find.

The next phase was to reconcile the two groups. Simply bringing them together to watch movies and other activities had no effect— the name-calling and skirmishes continued. Only when the researchers told the groups that the camp's water supply had been vandalized (suggesting the existence of a truly outside group), and that fixing it would require the help of both groups, did the two come together.

The intense competition between the Eagles and the Rattlers and their intense dislike for one another was due to the winner-take-all contests the researchers placed them in. But even in the absence of explicit competition, people still seem to favor those within their group over those outside it. However, adding competition between groups is a recipe for increasing cooperation within a group. This pattern appears to be especially true of men. In the Public Goods games, for example, men tend to cooperate much less than women do when the game is played among people in a single group. When the game becomes competitive against another team, however, men become hyper-cooperative within their team.[19] That is, they cooperate to compete against other groups. There is also evidence that people in higher-status groups tend to have overall higher levels of social dominance orientation—that is, they voice higher support for social inequality. But these attitudes aren't static. People become more willing to subordinate an out-group, and develop increasingly negative attitudes toward it, when that group is presented as a threat to their group's welfare.

The MIT political scientist Roger Petersen has examined how ethnic conflicts, such as those in central and eastern Europe, result from sudden changes in ethnic status hierarchies—when a group finds itself newly threatened by another.[20] This fuels the rebel instinct and drives status resentment and rage, emphasizing that these conflicts are driven as much if not more by emotion than by material interests. Indeed, one of the most surprising findings of Petersen's research is that historical hatred is often not the reason for such ethnic conflicts. Instead, groups typically target a group that they feel has illegitimately gained status over them, but is still close enough on the hierarchy that they feel they can turn things around and subordinate them.[21]

The intensity of our Status Dilemmas depends on the absolute amount of status available and its allocation—and on people's perceptions of who gets how much. A sure way to invoke group conflicts is to frame an issue as an intergroup struggle over a scarce, fixed resource. For example, this is how most immigration debates in the United States are framed today. Politicians who say an immigrant takes a job away from an American speak as though the number of jobs is fixed, so someone's gain is another's loss. To make the conflict even more intense, politicians may say that the number of jobs is shrinking due to a recession, outsourcing, or trade agreements.

Post-Scarcity Rebellion

Since absolute wealth and growth affect the intensity of status competition, the economic prosperity of postwar America played an important role in social and political change. Indeed, as the Harvard economist Benjamin Friedman has documented, economic growth drives political and social liberalization, including greater opportunity, tolerance of diversity, social mobility, fairness, and commitment to democracy.[22] By 1956, the United States had become the first society in which the majority of the labor force was employed in the service sector, indicating that a transition to Inglehart's post-materialist economy was already well under way.[23] Contrary to the idea that economic growth has little effect on happiness, in a twenty-five-year study of fifty-two countries between 1981 and 2007, Inglehart and his colleagues found that happiness increased in forty-five of those countries.[24] Central to this rising happiness was a growing sense of free choice, which was most predicted by economic growth and an increasing tolerance of diverse lifestyles. There is one practical sense in which economic growth is necessary to provide "ecological opportunities" for new lifestyles: consumer culture creates the material conditions necessary for diversifying social niches.

While both the Beats and the 1960s counterculture would oppose what they took to be the conformist materialism of their day, it's worth bearing in mind that both groups reflected a growing post-scarcity worldview. As the historian Christopher Gair notes, the baby boomers that would form the basis of the counterculture had considerably more freedom and disposable income than previous generations.[25] The enormous growth in college attendance also gave them more leisure time than earlier generations that entered the workforce right after high school. Many of the founding Beats met as students at Columbia University. Allen Ginsberg went there with plans to become a lawyer. William Burroughs had gone to Harvard and enjoyed a trust fund from his wealthy family.

The conditions for the emergence of rebel cool seemed to be set in postwar America, where issues of gender, racial, and other forms of equality were moving to the fore of social awareness, along with the rising economic power of its citizens. While these forces brought about increased economic and political participation, they also changed the structure of society itself by dismantling its hierarchical structure and its treatment of status as one-dimensional (synonymous with wealth). In terms of changes in consumption, we've seen that conspicuous consumption is driven by emulation—envy of those a bit higher in the hierarchy. The vertical dynamic of emulation requires three ingredients: a status hierarchy, an awareness of where you and others stand in that hierarchy, and a means to signal your position transparently to others. Status divergence, in contrast, requires its own motivations. It requires opposition to a group's norms about status, reduction or elimination of consensus regarding status rankings, and the means to signal distinct status norms. That's where opposition in the form of rebel cool enters the picture to undermine the three requirements of hierarchy we listed above. To look into these changes in more detail, we begin with the question: How do we measure status? Let's start with money.

The Status of Money

When we ask people if they think there's an intimate connection between wealth and status, they typically just stare at us with a look of consternation. Some suspect we've asked them a trick question. After all, it seems obvious that wealth and status are tightly linked. Many people reply by pointing out that socioeconomic status (SES) is such a powerful idea because it's the measure of social hierarchy in modern society. In other words, if you want to know your place in the social hierarchy, consult your SES. Despite the intuitive appeal of SES and its links to hierarchy, there are a lot of problems lurking behind this association. For one, the first "S" of SES is supposed to capture social facets of status, particularly educational attainment and occupational prestige. But there's less and less consensus over occupational prestige. For example, lots of people disagree over whether lawyers and CEOs have high or low occupational prestige.

Even leaving aside these questions, there are other glaring problems with using SES as a status ranking.[26] If SES were anything like status rank in chimpanzees, people would have to know where they fit. The fact is, most of us have no idea what our SES is. As the behavioral economists Dan Ariely and Michael I. Norton discovered, most of us have little idea how wealth is distributed in the United States.[27] This isn't altogether surprising. As the sociologist Randall Collins points out, our everyday experiences simply do not involve the rituals that once made social hierarchy visible.[28] For example, everyday experiences in traditionally hierarchical society were filled with gestures of deference—bowing, using formal modes of address, and so on. In England, touching the Queen, as Michelle Obama did in 2012, still causes an uproar because it's a violation of the codes of deference surrounding royalty. Obama's hug was much more in line with the American shift to reciprocal and egalitarian greetings.[29] As Collins notes, "The status order is invisible or visible only within specialized networks; occupation and wealth do not receive deference nor form visible status groups broadcasting categorical identities."[30]

Since people don't know their SES, researchers have been developing measures of people's subjective assessment of their SES.[31] The results are surprising. People's subjective SES depends on whom you ask them to compare themselves against: Are they better or worse off than all others in American society? Their neighbors? Others of the same race or ethnicity? Their parents at their age? In particular, there are revealing differences in the ways that whites and African Americans think about subjective social status.[32] Although traditional measures of SES suggest that African Americans are more economically disadvantaged than whites, African Americans have significantly higher subjective social status. Using subjective social status, African Americans feel they are doing better than members of their race/ethnicity, their neighbors, and their parents at their age. Indeed, more than forty years of research has confirmed that African Americans have significantly higher self-esteem than whites, and that it has increased dramatically compared to that of whites.[33]

This is where consumption is supposed to come into play as a way to advertise your SES to others. It was easy with sumptuary laws: just as a chimpanzee can glance at a fellow ape and know its status, sumptuary laws made a person's status easy to detect by simply glancing at a fur, silk, or coarse woolen garment they wore.

Conspicuous consumption took over when sumptuary laws fell out of favor to make some goods effective at signaling social rank. The wealth signaling that so troubled Veblen was the result of the extraordinary economic expansion of his age, driven especially by an explosion in manufacturing that expanded the economy by 400 percent between 1860 and 1900. With that sustained boom came the emergence of the nouveaux riches. For such an emerging class, the relation between a good and wealth could be straightforward. That is, an emulation good could signal a single dimension: wealth.

In a hierarchical society, emulation motivates consumption because we use consumption to improve our relative standing. For emulation to work, it means that some products—like traditional luxury goods—must transparently signal a social position. It also means that people must agree about what constitutes status. A luxury product, such as a Rolex watch or a Jaguar, for example, are only status symbols if status means wealth and people recognize the link between the product and its cost. Robert Frank suggests that the rise of the SUV corresponded to images linking the SUV to wealth, citing in particular Robert Altman's 1992 movie *The Player*, in which studio executive Griffin Mill, played by Tim Robbins, drives a Range Rover.[34] As these images strengthened the connection between SUVs and wealth, emulation, according to Frank, drove the increasing demand for them.

As it turns out, it is a lot trickier to signal status through conspicuous consumption than it may first appear.[35] In a mathematical model, it's easy for an economist to simply set someone's status in terms of their relative level of conspicuous consumption—the equivalent of a conspicuous-consumption 100-meter sprint. In other words, if there are ten people in the conspicuous-consumption race, the one who spends the most on conspicuous consumption wins the top status position, and so on all the way down to the last place. This gives an objective rank, but it doesn't factor in what other people think of one another, so it doesn't capture our intuition that status involves the esteem of others. For that to work in this context, we have to assume that everyone agrees to a social norm that determines a person's status by their level of conspicuous consumption. That's a pretty big assumption. Even during the Gilded Age, there was no such simple link. Two decades before Veblen's skewering portrayal of the leisure class in 1899, the so-called captains of industry had been decried as robber barons. The philanthropy of John D. Rockefeller, for example, was in part an effort to "fumigate his fortune," as the biographer Ron Chernow put it.[36]

There are no doubt groups that endorse the social norm that equates relative levels of conspicuous consumption with status. But trying to outconsume your neighbors or friends can also bring you disesteem. Consider the effects of relative house size on the cost of homes within a neighborhood. If the conspicuous-consumption norm is right, then homes that are larger than average in a neighborhood ought to fetch a premium. They offer a chance to stand a bit above one's neighbors by outconsuming them. In contrast, homes that are smaller than average ought to sell at a discount, since such homes are a sign of low relative position. A study of about six thousand homes in four hundred neighborhoods found the opposite pattern.[37] Larger-than-average-sized homes sold at an 11 percent discount, while smaller-than-average homes sold for about a 5 percent premium. It's possible that people are willing to incur some relative shame in exchange for getting into a better neighborhood and the benefits it provides, such as good schools. But why do larger-than-average homes not only fail to fetch a premium, but actually seem to be hurt by their larger-than-average size?

We suspect that the main reason larger-than-average-sized homes sell at a discount is due to the fact that people worry that owning such a home will make them the "neighorhood snob." In fact, people gain status within a group by being generous, developing a wide range of relationships with group members, and being committed to the group's well-being. They don't attain status by trying to dominate the group or trying to inflate their own worth.[38] So invoking envy and the rebel instinct in your neighbors by moving into a bigger house isn't a good way to win friends. It's also worth noting that the status of being a member of a neighborhood— collective self-esteem—doesn't depend on where one ranks within the group. The larger-than-average house violates the norms that build collective self-esteem, and may also bring disesteem because of these violations.

There's no simple direct link between one's level of conspicuous consumption and one's status. Indeed, this link may reflect mainly the early phases of economic wealth.[39] Over time, signals elaborate beyond a single measure of status as wealth. Consider Burgundy wines, Speyside single malts, modern art, and high fashion, in which goods signal not only wealth but discretion, refinement, and the capacity to navigate the complicated and murky rules underlying their use. Indeed, even beer, often thought of as an "everyman's" drink, now increasingly serves in this role, with the proliferation of microbreweries worldwide requiring the ability to distinguish between a Flemish Gruit Ale and a Czech pilsner. In the rapidly expanding Chinese economy, for example, Bordeaux wines were at first fairly straightforward signals of wealth, as the wines of a few leading producers, such as Lafite, were heavily valued. Although China remains the top importer of Bordeaux wines, more recently wine acquisitions in China are beginning to reflect the more multifaceted tastes of connoisseurship.[40]

Even in the seemingly straightforward case of conspicuous consumption as a signal of wealth, an emulation good may signal many social norms relating to social partner choice. Wealth signals extremely different qualities depending on whether it is inherited, accumulated as the result of obtaining a professional degree and working long hours to become a law partner or chief surgeon, or acquired through a questionable business venture. Goods can help disambiguate these sources of wealth; in some cases, they can also serve as signals of intelligence, cooperativeness and leadership, creativity, emotional stability, and the capacity to delay gratification. Consider, for example, a good as apparently regimented as a man's suit. A corporate lawyer's conservative gray Brooks Brothers suit signals very different traits than a designer's rakish plaid Etro suit does. Likewise, their offices will differ extensively in their furnishings—all intended to reinforce the valued qualities of each occupation.

Apparently straightforward signals of wealth, then, may signal many underlying qualities. Indeed, the elaboration of such signals is an integral component of costly signaling, as it makes a signal more powerful but more difficult to fake. The point is, the single-dimension signaling of status as wealth can easily fall into disesteem as more sophisticated signals emerge.

As evolutionary psychologists point out, displays of wealth signal an ability to provide resources for partners and offspring, and are thus a prominent signal involved in social selection. However, material wealth was probably not a major factor during most of human evolution and prehistory. The accumulation of material wealth was the exception rather than the rule. Other forms of wealth, particularly embodied wealth (such as strength and vitality) and relational wealth (social ties), were more prominent. We've seen that material wealth emerged relatively late, when societies shifted to production systems that were based on scarce defensible resources.[41] Therefore, as much as we may think the link between status and wealth is unconditional, it depends on underlying social norms.[42]

The Fragmentation of Status

Once we recognize the link between norms and status, we can see how alternative systems can emerge through connecting various norms to status. Sociologists originally studied this in the context of "deviant" subcultures, but consumerism clearly provides opportunities to integrate alternative status systems into the economy. Alongside the economic changes we alluded to above, the interplay between emulation and oppositional consumption diversified consumer lifestyles. Far from people thinking about a single status system in large-scale terms such as SES, status itself fragmented into more local statuses. Indeed, as the psychologist Cameron Anderson and his colleagues reveal, local status—the sort you get from the respect, admiration, and acceptance of your face-to-face groups—is much more important for your happiness than global socioeconomic status.[43] So far from being ranked in broad social structures, as consumer theorists traditionally assumed, status today is local.

Let's return for a moment to the dynamic of emulation and oppositional consumption. Table 2 summarizes some key themes that emerge from the diversification of status norms. Emulation, or envy, is the desire to have something—a possession or a trait—that someone else has. Opposition, in contrast, is the desire to have something that contradicts another group's status system. This contrary desire is invoked by the rebel instinct and the perception that the other group's system represents the status quo. In the absence of a strict hierarchy, the rebel instinct can be invoked by framing an

Status System	Consumer Motive	Psychological Reward	Self-level Impact	Signaling Good	Market Dynamic
Hierarchical	Emulation (status instinct)	Status rank	Individual self	Emulation good	Convergence
Oppositional	Hierarchical opposition (rebel instinct)	Cool (in-group respect, out-group disdain) Successful avoidance of negative association	Individual self	Oppositional good	Divergence
Pluralistic	Identification	Respect Successful avoidance of ambiguous signal	Relational/collective self	Identification good	In-group convergence/out-group disambiguation and divergence
	Out-group opposition	Cool (in-group respect, out-group disagreement) Successful avoidance of negative association	Relational/collective self	Oppositional good	

TABLE 2. Major categories of consumption. In pluralistic consumption, within-group dynamics may include emulation, but emulation across groups is less prevalent, since there is little consensus on across-group ordering.

out-group as mainstream simply on the basis of the market share of a product. Like consumer preference biases, the rebel instinct can influence consumer decisions without becoming a conscious element of deliberation. This is the essence of cool consumption. We'll examine two case studies below: the SUV and why it became popular so quickly, and an iconic advertisement from Apple. Both illustrate how oppositional consumption—not emulative consumption—impacts the brain's reward circuitry in a special way. This will introduce you to some key themes illustrated in the table on the facing page that emerge from the diversification of status norms.

SUVs, Relative Value, and Out-Groups

From the beginning of the SUV market segment in 1991, the Ford Explorer was the bestselling SUV on the market. Why was it so popular? Veblen would have said that it was because of emulation, but let's see if that's the whole story. Automobiles are obviously potent signals, and consumers face the challenge of trading off a car's functional characteristics (seating, fuel economy, safety) with a car's signaling properties. As SUVs were introduced in the early 1990s, they combined the social signaling properties associated with Jeeps—adventure, fun, at least metaphorically going off the road—with functional properties including seven- or nine-passenger seating and cargo capacity. As such, they represented a combination of functional and social signaling properties that hadn't been explored before in the auto industry. The rise of the Ford Explorer and of SUVs in general was partially due to the fact that the SUV segment expanded the spectrum of social signals people could communicate through their car purchases. But this diversification can't be fully explained without taking into account an oppositional dynamic.

To give you a sense of this dynamic, we have to go back to 1983, when an important event occurred in the U.S. auto industry—one that likely saved it. In late 1983, four years after its government bailout, Chrysler introduced the Dodge Caravan and Plymouth Voyager, whose sales were instrumental in rescuing Chrysler. The minivan segment grew to nearly 10 percent of the automobile market at its peak, but then something strange happened. People started associating the minivan with undesirable signals, which trumped its functional features. The minivan fell victim to uncool stigmas, culminating in the dreaded "soccer mom" and "daddy daycare" images.[44] As Toyota's national marketing director for trucks and minivans, Richard Bame, notes, "The stories we heard were, 'I just don't want to be seen in a minivan. I don't like being the soccer-mom joke or feeling like I've given up all trace of my identity to be a parent.' "[45] The concept of keeping up with the Joneses means we emulate them and so try to copy their purchases. But what happens if the Joneses own a minivan? In that case, the stigma of their minivan might actually increase the attractiveness of the SUV.

To get a better sense of relative value, consider some classic results from behavioral economics.[46] In 1992, Amos Tversky and Eldar Shafir asked three separate groups to suppose they were considering buying a new CD player. They pass by a store that's having a one-day clearance sale with a popular Sony player for $99, well below list price. Would they buy it? Now consider a variation. Beside the Sony player is a top-of-the-line Aiwa player for just $159, also well below list price. What would people do now? One final variation: Instead of the top-of-the-line Aiwa player, beside the Sony is an Aiwa model of inferior quality to the Sony but at around the same price as the Sony. Now what would they do?

The results were striking. In the case of the Sony only, 66 percent of participants responded they would buy the Sony (34 percent would hold off their decision to learn more about various models). When the Sony and top-of-the-line Aiwa are on sale together, 46 percent of participants said they would hold off their decision to learn more about various models. In this case, the conflict over which model to buy resulted in a 12 percent increase in putting off the purchase. But what's most interesting is that in the Sony and inferior Aiwa case, 74 percent of participants chose to buy the Sony without delay, up from 66 percent when the Sony is presented alone. In other words, comparing the Sony to the inferior product raised the attractiveness of the Sony.

Consider another classic case.[47] When given the choice between $6 cash and a nice Cross pen, 36 percent of participants chose the pen. But when another group was given a choice between $6 and the same Cross pen or a less attractive pen, 46 percent of subjects chose the Cross pen, a 10 percent increase. The less attractive pen is sometimes referred to as a decoy: the point of offering it isn't to provide an option people will actually take, but to make the Cross pen look even better and also to make the decision to choose the Cross pen simpler, since there's little conflict between the two pens. These sorts of experiments reveal that what we like isn't fixed and absolute. Instead, we often construct our preferences on the spot as we figure out what to choose. This means that factors such as available comparisons can have a major impact on what we choose.

Returning to the case of the SUVs, the value of SUVs came in part through comparing them with minivans. What's interesting is that the comparison isn't with functional properties—it's with the negative social image of an uncool group. The two car types overlap in functional terms, as families with children could see if they compared the two at the dealership. Minivans have even been shown to be safer than SUVs! So the rejection of the minivan was not on practical grounds. As we'll see, when products come with a social stigma—when a group views their signaling properties negatively— they can act as a repulsive force that "pushes" consumers toward an alternative product as much as, and maybe more than, the alternative product itself "pulls" them toward it.

We've mentioned this push-pull dynamic in the context of a direct showroom comparison, but there's another way comparative valuations can take place. That is, a minivan might also serve as a "reference point." Consider real estate. Whether a particular house is priced reasonably has little meaning out of context—the same house in different neighborhoods or regions of the country can have wildly different values. So you need a reference point. If you're in the market for a house, you might now go online to find the median price of similar homes in a neighborhood, or a comparable recent sale. An agent may prepare a comparative market analysis for a client, which they may claim is based on their own proprietary information, with the implication that their analysis provides a truer reference point. The choice of a reference point can have a dramatic effect, since the value of a house depends on how it compares to the reference point.

This can also happen in the case of cars. In fact, when you purchase a new car, your current car often serves as a reference point. If you are trading down, a less expensive new car may feel less valuable to you than if you are trading up for that car, because you're viewing the same new car from the perspective of the two different reference points—one higher and one lower. Steve is a case in point: Years ago, he traded in a Ford Windstar, a minivan, for a Ford Expedition, a lumbering SUV. Even though his children said the Expedition got its name from the trek they had to undertake to get to the seats in the back row, the most important thing about the Expedition was that it wasn't a minivan.

Going Negative

The sudden rise of the SUV depended in part on the fact that it entered the auto market as minivans were gaining their uncool stigma, which served as a sort of social repellent that drove people to SUVs. If comparing a Cross pen to one of inferior quality raises the value of the Cross pen, what happens in our brains when we compare a product to one that signals discordant or antagonistic norms? Can comparing a novel product like an SUV to a frumpy, quotidian one like a minivan trigger the rebel instinct? Neuroscientists have long wondered what happens in the brain when we successfully avoid a bad outcome—such as (some would say) when you leave the dealership driving the SUV instead of the minivan beside it in the showroom. The closest emotion to what you'd probably feel is relief. You typically feel relief when you finish doing something you find unpleasant, like filing your taxes or even unloading the dishwasher. But you can also feel it when you narrowly miss something bad, like just avoiding an accident when you're driving.[48]

To examine how the brain responds to narrowly avoiding something negative, the Caltech researchers Hackjin Kim, Shin Shimojo, and John O'Doherty combined fMRI with gambling tasks—some trials involved the chance to win $1 and other trials involved the chance to lose $1. The trials involving the chance to win money are traditional reward trials, while those involving losing money introduce avoiding an aversive event (losing $1). They found that on trials in which people successfully avoided losing money, the brain responded much as it did when people won money. Specifically, the ventromedial prefrontal cortex, the region of the brain we saw in chapter 2 that underlies the affective receipt of reward, activated in both cases. Conversely, the researchers saw decreased activity in the same region when people either didn't win in the reward trial or lost money in the avoidance trial.

There is now evidence that social rewards and avoidance of punishments elicit similar brain responses. The neuroscientist Gregor Kohls and his colleagues examined how the brain responds when people successfully avoid a social punishment.[49] To get a social reward—a video clip of an approving person—players had to press a button quickly after a square flashed on the screen. To successfully avoid a social punishment—a video clip of a disapproving person, who frowns contemptuously and gives the thumbs-down—players had to quickly press a button after a circle flashed on the screen. They found that both anticipating social reward and avoiding social punishment recruit similar activation in the ventral striatum, a structure that we know is critical in anticipating reward. They also found that players differed in whether they were more sensitive to punishment or to reward. This is reminiscent of the differences we saw between Lisa Ling and Alan Alda, explored in chapter 3.

These intriguing results reveal that the successful avoidance of a punishment has a similar hedonic impact to the anticipation or delivery of a reward, whether monetary or social, and activates the same reward circuits. Dodging a bullet, a close call, a narrow miss—all are treated as rewards by the brain. To put this into a real-life context, imagine you've arrived at a rental-car kiosk to pick up a car you've reserved, only to discover that the car you selected is unavailable and the only car left is one you don't like. Perhaps instead of a sedate family sedan like a Nissan Maxima, it's a menacing-looking Dodge Charger. As you begin to envision the embarrassment you'll feel driving that car, the rental-car agent realizes he's made a mistake and the car you wanted is in fact available. You breathe a sigh of relief. As you drive away, the relief you feel heightens the value of the car you're now in.

So if you choose an SUV, for example, after comparing it to a minivan you find uncool, you get the bonus of the rewarding effect of not choosing the minivan. It's an interesting emotion—we don't think there's a word for it. It's something along the lines of liking plus relief. As we'll see, this effect is one reason why cool consumption is a powerful force. We evaluate products such as SUVs relative to others we're considering, not only in terms of functional properties, as we would a pen or a CD player, but also in terms of their social signals. This suggests we should especially dislike products whose signals we associate with groups whose values aren't just different from ours, but antithetical to ours. Imagine a hippie having to drive a BMW, or a hipster in a Buick. In fact, consumers reserve their lowest ratings for such brands and goods.[50] What's particularly interesting about these negative evaluations is that they are typically the result of evoking our relational or collective self. In other words, they reflect in-group/out-group dynamics. The hippie doesn't want people to think he's a yuppie. The idea that consumption is about individual self-expression is pervasive.[51] Yet people are able to rapidly categorize various products—cars, clothing, music, and so on—by the social groups that use them.[52] The rise of "lifestyle brands," which have strong symbolic association with specific lifestyles, from the eco-consciousness of Patagonia to the surfer ethos of RVCA, is strong evidence that social consumption isn't primarily about individual distinction or uniqueness. It's about our group-level identity. As we've emphasized throughout this book, social selection, the profound groupishness it shapes in us, and our diversification into increasingly differentiated microcultures of consumption place a premium on signaling our group-level identity.

The prominence of our social identity in our consumption raises the importance of negative consumerism—that is, consumer preferences that depend not only on the positive evaluation of a good but on the avoidance of negative alternatives, particularly those that trigger associations to out-group identity. A consumer may choose to purchase an SUV in part out of identification with its social signals. But the purchase may also be made to avoid negative associations with out-groups, such as the stereotypical suburban family that would drive a minivan. Positive emotional evaluations (increased status) pull them toward the SUV, but avoiding the negative evaluation increases the rewarding effects of the purchase. It boosts the SUV's value by providing something like social relief (the successful avoidance of a punishment is in itself rewarding). So while two people who buy the same car are identical in their behavior, their rewarding experiences may be quite different, depending on the social comparisons at work in their brains.

So emulation motivates consumers only to the extent that products mirror a social hierarchy. Divergence motivates consumers to select products that oppose or don't fit into such a hierarchy. Divergence makes groups more and more different from one another. Take another look at the rise of the teenage rock-and-roll band. In forming bands, teenagers weren't trying to emulate athletes. What made rock and roll cool was its rejection of a dominant norm, which invoked the rebel instinct. Today, divergence among groups is driven by a variety of motives, only some of which are oppositional. Others may be simply a matter of disagreement or different interests, rather than the rejection of a dominant norm.

Opposition, in contrast, is the rejection—or at least the posturing at rejection—of a dominant status hierarchy. It would stretch credibility to claim that when SUVs first came out, by buying one instead of a minivan you were challenging the establishment. Yet the minivan became stigmatized by uncool norms, particularly staid images of domesticity, which are antithetical to cool (more on that in the next chapter). The SUV broadcasts its own positive social signals, such as freedom, adventure, power, and so on. But it also broadcasts oppositional social signals, such as the rejection of the staid domesticity of the minivan—offering an enticing consumption norm particularly to parents of young children who didn't want to be tagged with the uncool stigmas of the minivan. Kristen Howerton, who writes a blog titled *Rage Against the Minivan*, states, "It's just a symbol of women becoming the invisible, exchangeable mom—the soccer mom—where we all look the same and no longer have a sense of what's cool."[53] Even though automakers attempted to rescue the minivan's image through a series of ads in 2011 including heavy metal music and parents rapping about their "swagger wagon," minivan sales continue to decline, and some makers have left the segment entirely. Social signaling not only killed the minivan; it also created an oppositional dynamic that helped the SUV rise in popularity.

Oppositional consumption helped diversify status groups. This involved invoking the rebel instinct to oppose the values of the dominant status system in favor of alternative ones. Even as plural-istic consumerism has replaced hierarchical consumption, cool still requires at least the pretense of opposition. This may be as straight-forward as comparing a product against one that has the most mar-ket share, and so forming an unfavorable opinion of that product as too mainstream. Or you might choose a relevant out-group to op-pose, and so select products that compete with theirs. The out-group can even be an aging demographic that is becoming associated with a product, as we saw with Facebook. Harley-Davidson, the iconic rebellious good, is also aware of its need to continue attract-ing younger consumers as its customers' average age creeps up over fifty.[54] We were even approached by an iconic men's magazine, con-cerned that their brand's aging demographic meant they were los-ing their cool. For the rise of rebel cool, it is the oppositional dynamic between status systems—and the drive for identification and differentiation underlying this dynamic—that we need to look at to see what lies beyond emulation.

Apple: Emulation, Opposition, or Identification?

Consider the following use of a negative out-group and oppositional consumption. During the 1984 Super Bowl, an ad ran that is often regarded as the best commercial of all time, although it only ran once in prime time and the product it was advertising never even appeared in the ad. Directed by Ridley Scott, the ad featured a dys-topian world, almost postapocalyptic in its monochromatic dreari-ness. Human drones sit passively while Big Brother describes the Unification of Thought on a screen in front of them. Then an un-named heroine runs into the scene, wielding a sledgehammer. She is pursued by the Thought Police in riot gear. Just as Big Brother pro-claims victory, the heroine tosses her sledgehammer at the screen, causing it to explode. The screen fades to black and then the message appears: "On January 24th, Apple Computer will introduce Mac-intosh. And you'll see why 1984 won't be like *1984*."

Although Microsoft comes to mind, it was in fact IBM which was identified explicitly by Steve Jobs as Big Brother.[55] It's hard to think of a more evocative portrayal of a competitor as both a dissociative out-group—Orwell's droning proles, who may have lost the drive to rebel—and an oppressive hierarchical social structure led by Big Brother. Presented against this backdrop, the Macintosh isn't simply an emulation good. By making IBM such an oppressive, dominating reference point, the commercial positioned the Macintosh as an oppositional good, a good whose consumption was supposed to signal rejecting, opposing, and avoiding its competitor's negative associations. The link to rebellion is explicit, as the ad portrays an act of rebellion and invokes the rebel instinct directly.

The Apple ad also shows how strongly group-level identity figures in our consumption, how consumption is often more about signaling our group membership—our social niche—than our individual self. Indeed, Apple made group membership explicit in the 2006 "I'm a Mac" campaign. We think the reason such ads in 2006 emphasized group membership is because the forces of lifestyle diversification radically expanded the number of social niches. Group identity is now the principal consumer signal. That's one reason why branding has become so paramount: A brand acts like a secular symbol of group identity, sometimes referred to as a brand community. We'll call goods that primarily enhance a person's group identification identity goods. Many people's route to status is via membership in a group and conforming to its norms, rather than attempting to alter them. So rather than individual status rank, identity goods enhance respect (collective self-esteem).

The prominence of identity goods complicates the claim that consumerism is antithetical to community. Identity goods play a key role in both the fostering and maintenance of social groups.[56] Some prototypical identity goods are Harley-Davidson and Apple products. Within a status group, some individuals may compete for status by innovating: elaborating and complicating their group's norms, including consumption patterns, to both enhance their within-group status and to make the rituals of belonging hard for outsiders to copy. Think of such new goods added to a trendsetting member's repertoire as within-group emulation goods. A key difference between an emulation good and an identity good is that the value of an identity good doesn't decrease as more members of the group consume it.

Indeed, the value of such a good typically increases as more members within the group possess it. It becomes more emblematic of the group's common identity. As we'll examine, an identity good is devalued principally when members outside the group begin to use it. The value of Facebook among teenagers, for example, increased as more teenagers adopted it (indeed, it depended on a critical mass of such users) and only lost some luster when their parents started using it.

Emulation and identification are both driven by convergence dynamics, the tendency of individuals to converge on similar tastes or preferences. In psychology, these processes fall under the rubric of conformity, while in economics they are referred to as herd behavior. Identity goods, however, have a more complex psychology and involve divergence as well as convergence dynamics. Consumers should abandon identity goods when other groups start using them, because then the goods' signals become murky.

Marketing professors Jonah Berger and Chip Heath performed a series of intriguing experiments that supported the possibility that people abandon a good to avoid sending undesired identity signals to others.[57] In one experiment, they sold Livestrong yellow bracelets to residents of a Stanford University nonacademic dorm as part of a "wear yellow month" campaign. (They conducted the study before Armstrong's fall; the mass abandoning of Livestrong bracelets following Armstrong's confession is another powerful indication of their symbolic value and how such values can change.) A week later, they repeated this at the adjacent Academic Dorm, a dorm specifically requested for its extracurricular academic activities by students known as the campus geeks. A week after that, they examined how many members of the nonacademic dorm next door were still wearing their bracelets, and found a 32 percent drop (compared to only a 6 percent drop in a dorm across campus). The nonacademic dorm members didn't dislike the geeky students—it was just that wearing the bracelets could mean you were a member of either group. Berger and Heath suggest that people are most likely to stop using a product when a group very unlike their own starts using it. The standard competitive-consumption view predicts that people should be concerned most with what people most like them are consuming: for instance, a lawyer in San Francisco would care more about what cars others in the firm are driving than what local teachers drive. But that's not what happens. When dissimilar others adopt a product, its effectiveness as a signal breaks down. Divergence can be driven not by opposition to an out-group's norms, but simply by the need to keep your signals straight in a pluralistic setting.

Rethinking Keeping Up with the Joneses

The forces driving changes in consumer preferences, then, are more complex than we've considered before. As the number and diversity of status groups increase, signaling will intensify—not simply as the result of direct competition for status, but due to the need to separate each group's identity from the others. This is the principle of divergence at work. It requires a consumer culture that is creative in order to provide new opportunities for differentiation. This dynamic is a major force behind the transition from stable social hierarchy to pluralist microcultures as more and more unique social niches evolve over time. Consider virtually any consumer product category—clothing, music, cars—and there's likely to be an explosion of choice and a proliferation of niches. The average home now receives about 120 TV channels, and that's not including on-demand shows or other content providers, such as Netflix and Hulu. There are about 26 million songs on iTunes. And consider the proliferation of lifestyle activities. There are now so many kinds of yoga, including antigravity yoga, that a newspaper created a flowchart to help people decide what kind to take up.[58]

The implications of consumer lifestyle proliferation are large: the response to the Status Dilemma is the diversification of status groups. The multiplication of consumer microcultures doesn't mean more zero-sum competition. It means more routes to status are opening, and dissipating direct competition. It is the cultural equivalent of the adaptive radiation in nature that results in increased biological diversity.

Innovation—adopting some goods and abandoning others— becomes substantially more complex once we consider convergence and divergence motives alongside one another. People innovate within their group to gain status, but a lot of innovation comes from divergence across groups. This process fosters group identity and provides respect among the group (collective self-esteem). Sometimes this creates remarkably enduring identity goods. Consider the identity staple, the Bass Weejuns penny loafer, introduced first in the 1930s, worn by James Dean, John F. Kennedy, and Michael Jackson. When the preppy clothier J. Press attempted to update some of its clothes, its customer base howled in protest—since change is antithetical to the norms of tradition the brand represented. Consider also the Schott Perfecto leather motorcycle jacket, which recently celebrated its eighty-fifth anniversary as "America's signature bad boy uniform," as its makers call it. Worn by Marlon Brando in 1953's *The Wild One* and by James Dean, the Sex Pistols, the Ramones, Joan Jett, Jay Z, and Lady Gaga, its design is virtually unchanged. Speaking of that leather jacket, it's time to look deeper into rebel cool and its connections to the rise of oppositional consumption.

7

REBEL COOL

While Veblen wrote of the Gilded Age's emulation-driven consumption, a bohemian culture was rising in Greenwich Village, Montmartre, and other urban centers. Its avant-garde movements contained the seeds of the social, political, and artistic revolutions that would in short order reshape Veblen's Gilded Age. As the historian Peter Gay recounts, perhaps the most celebrated, if somewhat hackneyed, common theme of modernism—exemplified by such examples as Arnold Schoenberg's rejection of tonality in music, Wassily Kandinsky's exploration of abstraction in art, and James Joyce's stream-of-consciousness prose in literature—was its rejection of convention, the "lure of heresy."[1] Despite the avant-garde attack on convention and the bohemian attacks on bourgeois sensibilities, it wouldn't be until the 1950s that "cool" as a social norm would begin to emerge and its oppositional culture would become a visible rival to dominant status.

By the end of that decade, the broad contours of the first phase of cool—what we call rebel cool—had emerged out of the ethos of jazz musicians such as Thelonious Monk and Miles Davis, Beat writers such as Jack Kerouac and Allen Ginsberg, Hollywood idols such as James Dean and Marlon Brando, rock-and-roll stars such as Elvis Presley, and hipster essayists such as Norman Mailer. Rebel cool would appear as an oppositional norm, inverting the traditional

status system by appropriating the values and practices of those assigned to the bottom of the hierarchy: the African American, the
petty criminal, and the drifter. In particular, rebel cool meant conforming to social norms defined by five main elements: an oppositional style, an emotional style, a mode of experience, a sexual signal,
and a mode of masculinity.

Of these five elements, all but a mode of experience are social
signals, reinforcing the fact that cool follows the signaling logic of
social selection. Over the next thirty-five years or so, rebel cool would
become increasingly powerful as a transforming social force. Indeed, the intense signaling nature of cool is an important reason why
cool was so easy to assimilate into consumer culture. By the late
1950s, the assimilation was already well under way in the pages of
such lifestyle magazines as *Esquire* and *Playboy*. Responding to
what social critics perceived as a masculinity crisis, rebel cool offered a vision of masculine regeneration—a 1950s male rebellion
that preceded the 1960s sexual revolution.

Some accounts of this period view cool's integration into consumerism as a failure—what we call the co-option view of cool. In
contrast, we believe cool's greatest legacy was its transformation of
emulation consumption into oppositional consumption. For all the
retro-romanticism over rebel cool and the popular story of its fall
from innocence into consumerism, it's worth bearing in mind that
rebel cool had its share of problematic elements. In particular, it
embraced a clinical kind of rebellion and a thoroughly conservative
and troubling view of gender and sexuality. Far from being too radical for society to handle, its rebellious liberation was in places badly
misdirected and not nearly radical enough.

Thus, the best way to think about rebel cool is as a transitional
force. It unleashed social changes that would lead to a second, more
positive phase of cool. What we call DotCool began to emerge in
the 1990s as the diversifying forces of cool continued to reshape
society. There's no single watershed event that marks the transformation from rebel to DotCool—the latter's emergence was
gradual and depended on many parallel threads. In retrospect,
however, there were some signs that cool was undergoing a transformation in the early 1990s. Then–presidential candidate Bill Clinton's

1992 appearance on *The Arsenio Hall Show* wearing Wayfarers and beating out "Heartbreak Hotel" on his saxophone was one. The 1994 suicide of grunge rocker Kurt Cobain was another, as was the release of Netscape's Navigator web browser that same year. These were signs that a shift from hierarchical status to pluralist status was under way, more diverse lifestyles were proliferating, the earnestness of the oppositional stance of rebel cool was declining, and the ironic posturing of DotCool was rising. Spurred by the success of social movements, demographic changes, the growth of the Internet, and the fragmentation of the mainstream, the notion of hierarchy and the earnestness of the oppositional stance declined in the increasingly noisy and chaotic world of status cultures of the 1990s. Although DotCool retained aspects of rebel cool's oppositional stance, by now this stance also invoked posture and pretense. This is often taken to mean the rise of the frequently maligned ironic hipster. But, as we explore in the next chapter, these changes point to a more positive transformation of cool.

Both the rise of rebel cool and the transition to DotCool were responses to new economic realities, albeit starkly different ones. Rebel cool rejected the primary occupational metaphor of the 1950s: the man in the gray flannel suit, who drones along in an undifferentiated workplace, follows the rules of a creativity-killing bureaucracy, and marches in lockstep with other workers to someone else's drum. With the rise of a postindustrial "knowledge economy," rebel cool's rebellion became DotCool's unconventionality, and now drives innovation. Thus, the new occupational metaphor became the kid in the gray hoodie, who does something cool for a living like work in new media. The difference between James Dean and the actors Michael Cera and Ellen Page as icons of their generations captures this stark contrast. In this regard, cool now signals the learning that drives an innovation economy instead of the rebellion that rejects society altogether.

To take a closer look at the rise of the first phase of cool, let's examine its five major facets against the backdrop of Norman Mailer's 1957 essay "The White Negro," a defining if problematic manifesto of rebel cool, and the iconic movie of the day, *Rebel Without a Cause*. It's worth keeping in mind that Mailer's essay was in part his

self-conscious attempt to revive his career. He wrote it as he ago-
nized over whether he could ever replicate the early success he'd
enjoyed a decade earlier with the publication of his debut novel, *The
Naked and the Dead*. Based on his military experience in the Pa-
cific during World War II, *The Naked and the Dead* was an enormous critical and commercial triumph, spending more than a year
on the bestseller list and bringing him international renown.

The years leading up to the publication of "The White Negro" in
the quarterly *Dissent* were increasingly turbulent for Mailer, whose
pugnacious nature, supreme ego and insecurity, hard living, increasing drug use, and womanizing made it easy for him to alienate
friends, lovers, and wives. His novels *Barbary Shore* (1951) and *The
Deer Park* (1955) were disappointments compared to *The Naked
and the Dead*. *The Deer Park* was rejected by seven publishers before Mailer found a house that would take it on, making him doubt
the novel and abandon the genre for a decade. "The White Negro"
would bring him renewed fame and catapult him back into the spot-
light as a spokesman of hip.

It wasn't the first such essay. In fact, Anatole Broyard had written "A Portrait of the Hipster" a decade earlier for *Partisan Review*.
But Mailer took far more liberties with the realities of the Greenwich Village hipster and instead focused on the hipster as a white
middle-class male. Mailer's essay focused on the Beats as a lifestyle
more than a literary movement, offering influential images of new
gender roles that frequently bore little resemblance to those portrayed in Beat texts. Mailer's essay is in parts grotesque for its crude
racial stereotyping, cheesy in its now-outdated use of hipster lingo,
intentionally salacious and off-the-wall; it is also one of the most
anthologized essays of the twentieth century. "The White Negro" is
eerily prescient in crystallizing many of the strands of rebel cool
and in foreshadowing the social changes just over the horizon.

Cool as Opposition

Mailer's essay confronts the reader with a stark choice: either embrace the rebelliousness of the hipster or die a slow death by conformity. Consider the following passage:

> One is Hip or one is Square (the alternative which each new
> generation coming into American life is beginning to feel) [,]
> one is a rebel or one conforms, one is a frontiersman in the
> Wild West of American night life, or else a Square cell, trapped
> in the totalitarian tissues of American society, doomed willy-
> nilly to conform if one is to succeed.[2]

The rebelliousness Mailer has in mind is no run-of-the-mill dissent.
His model of the rebel is the psychopath, the forerunner of a new
kind of personality Mailer predicted would come to dominate society. He is a radical nonconformist who is violent, a sexual outlaw,
amoral, and unencumbered by conscience. This individual represented a union of the bohemian, the juvenile delinquent, and the
African American of the jazz scene Mailer frequented in New York,
for "the Negro was forced into the position of exploring all those
moral wildernesses of civilized life which the Square automatically
condemns as delinquent or evil or immature or morbid or self-
destructive or corrupt."[3] The hipster was an American representation of the existential hero, who rebelled against the middle-class
values of ambition and materialism by embracing the spontaneous,
carefree, drug- and sex-filled world of the "Negro."

This was no simple bohemian rejection of bourgeois conventionality. This opposition was rooted in the far more subversive notion that civilization itself was a sort of madness—that the mass
society of the 1950s was creating a particularly dangerous mental
affliction—and the only way to avoid being infected was to rebel
against it. The first element of this view traces to what is among the
most historically pernicious works ever penned, Freud's 1929 *Civilization and Its Discontents*. Freud's basic view was that human
nature—rooted in destructive, violent, and sexual drives—is intrinsically antisocial. In Freud's words, "Generally speaking, our civilization is built up on the suppression of instincts. Each individual has
surrendered some part of his possessions, some part of the sense of
omnipotence or of the aggressive or vindictive inclinations in his
personality."[4] Civilization is a coercive force designed to control
these drives. It can't, however, dissolve these drives; it can merely
redirect them. Through sublimation, for example, it can redirect
sexual energy into a socially acceptable form—namely, productive

work. Civilization can also repress, and frustrate, these drives. But the libidinal energy has to be released somehow. It finds its convoluted way out through narcissistic "neuroses" of megalomania or depression, psychosomatic illnesses, phobias, obsessive and compulsive behaviors, and so on. Freud's views are particularly damaging with regard to many mental illnesses, which he believed were rooted in "complexes" caused by repressed sexual and aggressive wishes from childhood (a view that has been largely discredited).[5]

With the rise of fascism and World War II, the Freudian view of civilization as the source of neuroses spurred a cottage industry of psychoanalytic critiques of mass society. An influential early work in this tradition was Erich Fromm's 1941 *Escape from Freedom*. Fromm argued that people escape the responsibility and uncertainty of freedom by submitting to authoritarian rule, thereby conforming to and imposing that order on others (reminiscent of social dominance orientation). By the 1950s, more and more social critics began to sign on to the idea that mass society created a deficient personality type. One strand of this diagnosis, illustrated by William Whyte's 1956 *The Organization Man*, reflected the sociological critique that mass society created a deadening conformity. In other words, society creates needs, tastes, desires, and preferences to drive the system of mass production and, in turn, mass consumption. Whether authentic or manufactured, the result is the same: emulation leads to imitation, a mass desire for the same things. Mass consumption is conformist consumption: bland, homogeneous, and dreary. (The logic of mass consumption requires this conformity, so the argument went, summed up in Ford's quip that you could choose any color of his Model T so long as it was black.)

The rise of the Cold War also made for uneasy comparisons between mass society and totalitarian society. The rugged individual seemed a threat to both—and both required oppressive controls to keep individuality in check. These critiques flourished despite the fact that this period was one of dramatic economic growth that led to an unprecedented standard of living. In fact, affluence itself became a problem, and especially the question of whether America had essentially traded away its soul for material comfort. You can find strands of this view from critiques of mass society and conformity in the 1950s to recent critiques that consumerism depends on infantilizing our tastes and fears that globalization will reduce us all to indistinguishable generic consumers.

In a nod to Freud, Mailer's basic argument was that the pace of modern society so overstresses the nervous system that sublimation no longer works to make adjusted personalities. Since conformity via sublimation is no longer possible, conformity can only work by killing our inner drives. So conformity in the mass society of the 1950s ages and dulls the spirit, meaning an individual "conform[s] to what he loathes because he no longer has the passion to feel loathing so intensely."[6] Conformity defeats the spirit and turns humans into passionless, droning automata. While these views had become popular themes of his day, Mailer's next move cemented his essay as a bizarre but widely influential statement of rebel cool.

Mailer asked: What must one do to avoid a slow death by conformity? He found his answer in the curious criminal psychologist, hypnotherapist, and author Robert Lindner. Lindner, who was good friends with Mailer, was a rising star among the social critics of the 1950s when he died suddenly of heart failure at the age of forty-one. Now largely forgotten, Lindner was the author of the 1944 book *Rebel Without a Cause*. Despite its title, the book bears little resemblance to the James Dean film—it is a study of psychopathy based on hypnotherapy sessions Lindner conducted with a prison inmate. By the 1950s, Lindner had grown fundamentally suspicious of psychological and psychoanalytic emphases on adjustment—that the goal of therapeutic practices was to produce individuals who were well-adjusted to society. Calling adjustment the single greatest myth of his time, Lindner believed that psychologists and psychiatrists had become tools of a totalitarian-like status quo.[7] The ideology of psychological adjustment was creating a conformist culture that was also unintentionally creating psychopaths in unprecedented numbers. In two popular books, *Prescription for Rebellion* (1952) and *Must You Conform?* (1955), Lindner argued that the root cause of these social ills was the fact that we have an "instinct for rebellion" that was smothered by conformist mass society and by a psychoanalytic industry turning people into social cogs.

Lindner distinguished between a positive and a negative form of rebellion. As the sociologist Paul Halmos paraphrased Lindner's thought in 1957, "Positive rebellion is the creative adventure of the individual beyond the boundaries of the conventional, the routine and the stagnating; negative rebellion is destructive, anti-social and arrests development."[8] What worried Lindner most was his belief that conformist mass society had so stifled positive rebellion that it was creating a generation of negative rebels: the psychopath and the juvenile delinquent. His worries were featured in major publications, such as *Time*, which ran a disturbing article in 1954 titled "Rebels or Psychopaths?" that dealt with the growing issue of juvenile delinquency from Lindner's perspective. By that time, Lindner's concerns had only increased: "It is not youth alone that has succumbed to psychopathy, but nations, populations—indeed, the whole of mankind. The world, in short, has run amuck."[9]

Mailer frequently visited Lindner in Baltimore, where Lindner was working in various psychiatric facilities. In their conversations, Lindner taught Mailer why expressing the rebellious instinct was vital for mental health, and how the link between violence and conformist culture was creating a psychopathic society. Yet Lindner's positive rebellion got lost somewhere in translation. It turns out to be the act of a "mature" rebel, and Lindner cites the Founding Fathers as prototypical examples and the Bill of Rights as a product of useful rebellion.[10] Mailer likely knew that such a familiar and reassuring image of the rebel was far too conventional for his hipster manifesto. Instead, as the new rebel, Mailer exalted the psychopath—a term that appears almost fifty times in his essay. Positive rebellion now out of the picture, Mailer turned the psychopath's uninhibited pursuit of sex and violence into the rebel's highest ends—acts of an American existential hero. For Mailer, the psychopath was a new kind of personality that would come to dominate society by refusing conformity and sublimation.

Mailer's enthusiasm for psychopathic rebellion raises a critical distinction. As a rebel without a cause, the rebel in a psychotherapeutic context bears little resemblance to the rebel who protests against injustice. The latter had also come to the fore of public consciousness in the 1950s, particularly with the Nobel Prize–winning novelist and essayist Albert Camus's *The Rebel* in 1951. For Camus, the act of rebellion creates values, dignity, and solidarity (Camus says, "I rebel, therefore we exist"). Mailer's view, in stark contrast, is essentially Lindner's negative rebellion.

Today, a growing neuroscientific approach to psychiatry studies Lindner's negative rebellion, which is still referred to at times as psychopathy and at times as antisocial personality disorder, characterized by low empathy, blunted emotion, reactive anger, and disregard for social norms. Men are eight times as likely to be diagnosed with antisocial personality disorder as are women.[11] Psychologists also study rebelliousness, which they describe as risky behavior that has poor health outcomes.[12] For example, some people respond with "reactive rebelliousness" when they are told to follow a rule or comply with a request, such as when a parking attendant tells them they can't park their car in a particular location. These requests may be answered with retaliatory vengefulness. There's no reason not to follow the rule; these people just don't want to be told what to do. A related clinical form of rebelliousness is "proactive rebelliousness," which refers to people who get a thrill from breaking rules.

For this reason, rebel cool as Mailer envisioned it didn't have its roots in a political stance. Indeed, the Beats, with the notable exception of Ginsberg, were for the most part ardently apolitical (by this time, the fact that the Beats might have expressed rebelliousness through their artistic creativity was largely ignored). The whole point of the oppositional stance for rebel cool was now to stand outside of society, to oppose its domesticating bonds—not to change it, but to drop *out* of it.

Some of the Beats followed a trajectory like that of William Burroughs, author of *Naked Lunch*. Burroughs came from a well-to-do family, attended Harvard, and received income from a trust fund—yet embraced a lifestyle of drugs and violence. He accidentally shot and killed his wife in a drunken game of William Tell in 1951, and spent years struggling with heroin addiction. And Neal Cassady, who was the inspiration for Dean Moriarty in Jack Kerouac's *On the Road*, was perhaps the one who most embraced the Beat lifestyle. Raised by an alcoholic father, Cassady spent his life involved in petty crime, sometimes homeless and in and out of jail. He wandered with the Merry Pranksters until his death in Mexico at the age of forty-one, reportedly from an overdose of secobarbital.

Cool Experience

To Mailer, conformity was a slow death because its civilizing pressures extinguished unconscious, primal drives—the psychic energy that is the source of life. To "be with it" is to have experiences that put one in touch with these primal drives—"energy, life, sex, force"—a state Mailer compares to the mystic's grace, the "paradise of limitless energy and perception." This is another reason why the psychopath is so prominent in Mailer's essay. According to Lindner, the defining quality of the psychopath was his inability to delay gratification, acting impulsively and instinctually. This made the psychopath childlike in his present-mindedness. It also made him violent and hypersexual. So too, Mailer argues, the hipster lives in the "burning consciousness of the present," where he can "remain in life only by engaging death." To be alive, then, is to experience the sense of the body.

Among the most influential work in this tradition was Herbert Marcuse's *Eros and Civilization*, published in 1955. Whereas Freud saw society as necessarily repressive, Marcuse believed in the possibility of a non-repressive society, which would involve the liberation of erotic instinct, or Eros. Marcuse believed that mass society depended on the repression of Eros for its productivity. The revolutionary liberation of Eros would destroy repressive social institutions. Marcuse believed that technology could free man from labor, the sort of work that is alienating, and allow for a new form of creative work as play. Real work, he thought, would liberate Eros in a somewhat murky idea he called polymorphous eroticism. While we're not entirely sure what liberated Eros work would look like, unleashing libidinal energy was a central tenet of the 1960s belief in the revolutionary powers of "love," expressed both in the free love movement and in the sexual radicalism of Mailer's essay.

Cool as Emotional Style

The promotional poster for *Rebel Without a Cause* shows James Dean leaning casually against a wall, one foot crossed over the other, hand in pocket. His head tilted slightly, he's staring straight into the camera with a challenging expression that recognizes your presence but intimates he's not fazed by it. It's the sort of pose a fighter might take to belittle his opponent—hardly giving him the time of day. It would embody the Hollywood face of cool and rebel cool's emotional style—aloof, detached, and above all, in control. At age twenty-four, Dean would die in a car crash a few weeks before *Rebel Without a Cause* opened in 1955, cementing his image as the too-fast rebel.

Dean's pose would become the face of cool. But the oppositional stance that emerged in the 1950s could have taken any number of emotional styles. It wouldn't have been a bad bet to predict that emotional liberation and freer, more open expression would emerge as dominant emotional styles. The psychoanalytic critiques of mass society that helped fuel the impetus to rebel centered on how mass society's technocratic rationalism was creating numbing conformity and people unable to feel genuine emotional connections to one another. Many critics, such as the humanistic psychologist Abraham Maslow, warned that genuine relationships were being replaced by our relationship with material goods. This would lead to the human potential movement of the 1960s, with its "encounter groups" where people would "let it all hang out" to get in touch with their authentic feelings. Even earlier, J. D. Salinger's *The Catcher in the Rye*, published in 1951, would resonate with a generation of adolescents who felt emotionally alienated from their stoical, inexpressive Depression parents, who seemed concerned only with material security. This would bring the first rumblings of the "generation gap" into public consciousness and make emotional detachment between the generations a major social concern. In the 1950s, Lindner and others were already blaming the inability of parents to connect with their children as the root cause of growing juvenile delinquency. But emotional blossoming wasn't to be the way of Mailer's hipster.

The emotional disconnect between generations and the dysfunctional emotional life of the older generation became the central themes of the extremely popular family melodramas of the day, such as *Rebel Without a Cause*. Jim (James Dean's character) was a "delinquent"—the product of an overbearing mother and a meek father, who provide a comfortable lifestyle but are unable to form genuine emotional relationships with their son. Likewise, Judy, played by Natalie Wood, is unable to find an emotional connection with her father, despite her attempts to do so. When she tries to kiss him as the family is sitting down to dinner, he becomes so flustered that he slaps and scolds her. Jim and Judy are the quintessential 1950s teenagers, emotionally cut off from parents whose materialism has deadened their emotions. The teens both turn to delinquency in search of the family love they can't find at home, with Freudian complexes pushing them along. Dean's character is explosive in his interactions with his parents (the film is a melodrama, after all). Among the most iconic of such scenes is Jim's anguished cry, "You're tearing me apart," competing with Marlon Brando's "Stella!" as one of the screen's most memorable tormented emotional outpourings.

While Jim's turmoil is vital in portraying the generation gap, the emotional style of rebel cool was portrayed through Jim's interactions with his peers. Jim may have lost control, or "flipped out," as Mailer's hipster would say, in front of his parents, but respect among his peers demanded keeping his cool, especially when his manhood was challenged. In particular, the emotion to really restrain and keep hidden was fear. According to Mailer, flipping, or losing control by showing fear, was the worst transgression the hipster could commit. Nothing could set Jim off like the insult of being called chicken, and his fights over the insult were the reason the family kept moving from town to town. Indeed, the central plot device of the movie, the chickie game, is racing cars toward a cliff to see who will bail out first, a contest over who can keep his cool the longest.

Jim's turmoil around his parents and his cool around his peers illustrate an important element of rebel cool. It isn't simply about not having emotions, despite Mailer's fascination with the psychopath. We need to see Jim's emotions—it's his ability to control them and hide them from his peers that makes for rebel cool. That is, rebel cool prescribes a social norm around emotional expression. This norm has its roots in at least three sources. One source traces cool emotion to African American culture, where it emerged as a strategy during slavery for negotiating interracial interactions. It became a signaling mechanism to convey power, detachment, and style.[13] A second source traces it to an emotional style that emerged between the 1920s and 1950s in Protestant American middle-class culture.[14] This style was evident in the arts as abstraction and formalism replaced concrete representation and sentiment. In the music of Mailer's hipster, the cool jazz of Gil Evans, Dave Brubeck, Miles Davis, and others was growing in popularity against the more emotionally expressive bebop of Charlie Parker and Dizzy Gillespie. The third source roots the cool emotional style in a perceived crisis surrounding masculinity, which we turn to next.

Mode of Masculinity

The 1950s are often remembered for their rigid female gender roles, and in particular the stereotype of the idealized suburban housewife as represented on television by Barbara Billingsley (as June Cleaver), Donna Reed, and Harriet Nelson. But Ann Doran's portrayal of Mrs. Carol Stark in *Rebel Without a Cause*—so unlike the image of suburban domestic tranquillity—illustrates the deep contradictions and tensions surrounding the 1950s housewife. These contradictions were due in no small measure to the tragic influence of Freud's view of women. On the one hand, Freud's theory of women served as "scientific" justification for the 1950s status quo despite the fact that this theory was nothing more than his own fantasy, with no scientific support (because it was intrinsically unscientific).[15] Freud viewed women as naturally and morally inferior—girls were essentially failed boys and women were castrated men, their inner life defined by penis envy. Women were sexually passive because there is no feminine libido, according to Freud, so women only have sex to have children—and they only have children, according to Freud, because a child substitutes for a penis. Freud dismissed a woman's desire for equality as a neurosis also rooted in penis envy and wanting to be a man. Hence, being a housewife and caring for children represented, for Freud, the only source of female psychic fulfillment. Betty Friedan, following Simone de Beauvoir's *The Second Sex*, would expose the role of Freud's view of women by outlining what she called the feminine mystique. However, Freud's theory also placed new focus on the problematic role of the mother in a child's psychological development. In postwar America, this led to a new national menace—none other than moms.

Carol Stark in *Rebel Without a Cause* thus portrays a force that precipitated a crisis in American masculinity. "Appropriate" mothering supposedly required a mother to loosen her grip of maternal love as a boy develops, which would allow a son to "resolve his Oedipus complex." That meant he would become a masculine heterosexual—the goal of adjustment psychoanalysis. But growing numbers of mothers were supposedly doing just the opposite by dominating and smothering their sons. The most vitriolic attack on mothers came from Philip Wylie, whose *Generation of Vipers* introduced America to the term "momism." First published in 1942, it would become a runaway bestseller—and by 1954 it was in its twentieth printing. According to Wylie, "The Oedipus complex had become a social fiat and a dominant neurosis in our land."[16] In his view, moms, "spiritual saboteurs," had taken over the land.

Wylie would claim that his screed was sarcastic, but what followed would be anything but a joke, as momism became a national anxiety. Among the most prominent voices against moms was Edward Strecker, chair of the Department of Psychiatry at the University of Pennsylvania. His 1946 book, *Their Mothers' Sons*, was excerpted in *The Saturday Evening Post* and featured prominently in other major magazines. Strecker argued that more than 1 million men had either been rejected from military service or discharged during the war because of psychiatric disorders that were rooted in overprotective moms. Strecker even speculated that Hitler, Mussolini, and Stalin had all been too attached to their mothers.

Strecker wasn't finished. He went on to blame momism for homosexuality and communism—the so-called lavender and red menaces. Homosexuality and communism were frequently linked, particularly during the 1950s McCarthy-era witch hunts. Consider, for example, Reverend Billy Graham's approval of McCarthy and other Capitol Hill inquisitors for "exposing the pinks [liberals], the lavenders, and the reds who have sought refuge beneath the wings of the American eagle."[17] Strecker's link between momism and communism came from the idea that men who are the children of these moms go on to create "feminized" social structures. Rather than being individuals, they seek the security of groups that will protect them. (The link between momism and homosexuality wasn't simply a fringe idea. Even Betty Friedan made the connection in *The Feminine Mystique*, and cited the negative consequences of momism as further grounds for her case against the American housewife.)

The 1948 publication of the Alfred Kinsey report *Sexual Behavior in the Human Male* also raised new questions about male sexuality. The book was a surprise runaway hit, and Kinsey's findings regarding male homosexuality were among the most discussed and debated parts of the book. Kinsey claimed that 37 percent of males had at least some overt homosexual experience to orgasm, and 10 percent of males were more or less exclusively homosexual. Kinsey also stated that "males do not represent two discrete populations, heterosexual and homosexual." Instead, he developed a seven-point scale from exclusively heterosexual to exclusively homosexual and stated that males fall somewhere along a continuum. Christine Jorgensen, who had served in World War II as George Jorgensen, also became an instant celebrity in 1952 when the *New York Daily News* published a front-page story, "Ex-GI Becomes Blonde Bombshell," about the first widely known American transsexual to have undergone sex reassignment surgery. This, too, led to growing anxiety over traditional notions of sexual identity and gender.

These themes are played out in *Rebel Without a Cause* through the character of John "Plato" Crawford, portrayed by Sal Mineo. The allusions to Plato's homosexuality were subtle, but not subtle enough to escape a Production Code officer, responsible for making sure industry moral censorship guidelines were followed. He sent the film executive Jack Warner a memo warning against any implied homosexual relationship between Jim and Plato (in real life, the director Nicholas Ray and James Dean were bisexual, and Sal Mineo was gay).

Ann Doran's character in *Rebel Without a Cause* was not only causing anxiety in her son; she had also emasculated her husband. Frank Stark, played by Jim Backus, is shown in key scenes wearing a frilly apron and in another hurrying to clean up a mess before his wife finds out. At one point in the movie, Jim becomes so frustrated by his father's inability to tell him what a man is that Jim attacks him. Frank Stark thus reflects the growing anxiety both that the suburban house had become matriarchal and that the Organization Man had become domesticated and emasculated. A remarkable illustration of these anxieties is Arthur Schlesinger Jr.'s 1958 essay in *Esquire*, "The Crisis of American Masculinity." In it, he blamed the crisis on men spending too much time huddled in groups and faceless organizations, saying, "One of the most sinister of present-day doctrines is that of *togetherness*."[18] While he didn't blame the crisis directly on Mom, he said the group gave men a "womblike security," again equating the group with the feminine. In short, conformity and the group now appeared suspiciously "unmasculine." The sociologist David Riesman even suggested in *The Lonely Crowd* that male sexual performance anxiety was due to the 1950s wife, saying, "The relatively unemancipated wife and socially inferior mistresses of the inner-directed man could not seriously challenge the quality of his sexual performance."[19]

So the Beats appeared as a vision of masculine regeneration. They left domesticity behind to head out on the road, where the only women they encountered were potential conquests, and secondary to their nomadic life. As the English professor Susan Fraiman notes, cool became a mode of masculinity defined by a rejection of the maternal that flouts conventionality but embraces a conventional view of gender at the same time. In her words, "The project of [the Beats'] circle was to glamorize a nomadic male camaraderie in principled flight from women and their supposed conventionality."[20] The Beats themselves echoed a remark Freud had made to Wilhelm Fliess in 1901: "In my life, as you know, woman has never replaced the comrade, the friend."[21] Indeed, Freud's wife, Martha, a friend noted in her journal, was resigned to the fact that Fliess could "give her husband something beyond what she could."[22] When Freud fainted in a Munich hotel a few years later, he blamed it on the fact that he had been there earlier with Fliess and it had conjured up in him an "unruly homosexual feeling."[23]

Ginsberg said that "the social organization which is most true of itself to the artist is the boy gang."[24] Mailer made the connection between rebel cool and hypermasculinity directly when he said that to show fear meant "to reveal the buried weaker more feminine part of your nature." Burroughs was more blunt: He said that women might well be a "biological mistake."[25] Alluding to Burroughs's wife, Joan, Kerouac's biographer Ellis Amburn calls her "the best example of the Beats' abuse of women, a sacrifice on the altar of bogus masculinity and dishonest sexuality."[26]

Mailer's own writings became the target of feminist scorn, along with his incarnation of a bombastic machismo, particularly in the 1970s when he aimed his writing against Kate Millett's *Sexual Politics* in *The Prisoner of Sex*. The image of masculinity that Mailer portrays in "The White Negro," particularly the search for absolute sexual freedom and a quote from Lindner on the psychopath's need to rape, echoes a Freudian view of sexual aggression as natural, and makes it the goal of the hipster as he discards society's encumbering conscience. Today, this combination of traits that Mailer blends into his machismo is sometimes referred to as the "dark triad": Machiavellianism, narcissism, and psychopathology.[27]

Almost instantly after the publication of *On the Road*, magazines were focusing on its portrayal of the Beat lifestyle as a way to reclaim masculinity. *Playboy* and *Esquire* in particular couldn't get enough of Beat articles—Kerouac alone wrote more than a half dozen articles for them in the year following *On the Road*. In 1958, *Playboy* announced its readership was the youngest, best educated, and most affluent in the country—the leading edge of a new male. *Playboy*'s spread on "The Beat Mystique" that same year featured an article by Kerouac. But the Beats, the spread argued, were kind of a drag—penniless and nihilistic. The magazine presented a lifestyle that offered a way to keep Beat nonconformity while enjoying the finer things in life. Hugh Hefner dubbed them the Upbeat Generation, who "embrace the play and pleasure aspects of life along with the work. So we were rejecting the notion of conformity, but turning life into a celebration that incorporated capitalism."[28]

While *Playboy*'s presentation of the Beats' lifestyle offered a reassuringly conventional image of masculinity, *Playboy* also connected the Beats' free spirit to an image of straight-and-narrow sexual identity that was markedly different from theirs. Mailer acknowledged that hipsters are often bisexual, as were many of the icons of rebel cool, including Jack Kerouac, Neal Cassady, Marlon Brando, and James Dean.[29] However, this fact would soon be expunged from the public persona of rebel cool. Indeed, as Barbara Ehrenreich notes, Hugh Hefner added pictures of nude women to *Playboy* despite the fact that he was at heart puritanical, and the pictures were more antiseptic than erotic in their airbrushed artificiality. He did so to assure the magazine's readers that the new male single life it portrayed was thoroughly heterosexual.[30]

Cool as Sexual Strategy

Above all, Mailer's musings on sex made his essay scandalous and brought him the notoriety he sought. The distinction he draws between love as the search for a mate and the search for orgasm is central to the hipster image: "The drama of the psychopath is that he seeks love. Not love as the search for a mate, but love as the search for an orgasm more apocalyptic than the one which preceded it."

The Beats were also enthralled by the whole idea of apocalyptic orgasms, which they and Mailer learned from Wilhelm Reich, one of Freud's wackier disciples. Reich believed that neuroses stemmed from the inability to have an uninhibited orgasm, which released an invisible life force he called orgone. Reich also built an orgone accumulator, which was thought to capture the universe's orgone energy and heal all maladies and lead to better orgasms—Woody Allen's famous orgasmatron in *Sleeper* is a parody. Kerouac featured an orgone accumulator in *On the Road*, and Ginsberg, Mailer, and others all spent time in one, hoping to capture orgone energy.

Leaving aside the more mystical quest for orgasm, Mailer, the Beats, and Hefner were all foreshadowing the coming sexual revolution of the 1960s. While it's called a revolution, it's worth mentioning that the sexual revolution followed the broader historical trajectory of sexuality in the United States, from a means of procreation within the family to a birthright of pleasure and a critical aspect of personal happiness.[31] Indeed, Kinsey's early reports on male sexual behavior and in 1953 on female sexual behavior showed that what was going on behind closed doors was a lot different from the public profession of traditional "sex as procreation within marriage" attitudes. For example, Kinsey reported that two out of three males had premarital sex and about half of all females did too, though much of this premarital sex for both males and females was with the person they would eventually marry.

In some ways, then, the sexual revolution was about public attitudes catching up with private practices. Even though the sexual revolution got going in the 1960s, it's not as though people weren't already intensely curious about what other people were up to sexually. In 1953, only the Bible and a popular book on positive thinking sold more copies than Kinsey's report on female sexuality. That same year, Kinsey was featured on the cover of *Time*, and he quickly became a household name.

Technological and demographic changes also helped kick-start the sexual revolution. The birth control pill was approved in 1960, and within a year 2 million women were taking it. (By 1990, 80 percent of women born after 1945 had been on the pill at some point in their lives.)[32] People also are waiting longer to get married. The age at which people are getting married has increased from twenty for women in 1960 to twenty-five for women today. In 1960 it was twenty-three for men; today it is twenty-seven. These additional pre-marriage years are increasingly spent away from home at college. Enrollment more than doubled between 1960 and 2000. The number of women attending college tripled between 1960 and 1972, with women now outnumbering men on campus. These changes have had enormous social implications. The sexual revolution— like much of the 1960s counterculture—began on college campuses, where young adults experienced a newfound autonomy amid unprecedented levels of material comfort, which helped provide opportunities for new experiences, including buying records, traveling, and experimenting with drugs.[33]

The signaling style of rebel cool played a pivotal role in the sexual revolution as well. To see how, we're going to ask you to imagine looking at the faces of Leonardo DiCaprio in *Titanic* and Daniel Craig in his role as James Bond. If you prefer, you can think of Orlando Bloom and Arnold Schwarzenegger. Craig and Schwarzenegger have much more masculine, rugged faces than do DiCaprio and Bloom. People typically think that those with "cute" faces are more trustworthy. But there's more to it than that. Heterosexual women typically judge the cuter faces to be those of better potential long-term partners. In contrast, women typically judge the more rugged faces to be better potential short-term hookups.

These preferences reflect different sexual strategies people pursue—what we look for in a partner depending on the kind of relationship we are after, and especially the duration. For example, chances are that if you had to choose a sperm or egg donor you wouldn't choose randomly. You'd want the fertility clinic to tell you something about the donor. In particular, you'd likely want to know whether the donor was healthy, how tall they were, whether they were attractive, intelligent, and so on. Since your relationship with the donor would be limited to a single, anonymous transaction, you'd be essentially inquiring about their genes. These considerations are similar to short-term sexual strategies. Masculine features are thought to be unconscious signals of good genes. They are the product of high testosterone, which can only be borne by individuals with strong immune function.[34] Long-term sexual strategies, on the other hand, are ones in which both partners make a serious parental investment in child rearing.

Human sexual strategies are an extremely complex blend of short-term and long-term goals, which reflect both our evolutionary past and our social context. For example, in one study, female preferences for masculine male faces increased as the rate of disease increased across thirty countries.[35] These results illustrate that while evolution has played a role in structuring our preferences, we are also sensitive to context. Our choices typically reflect an adjustable trade-off between a mate with the qualities of a good long-term partner and one with good genes. While these are not mutually exclusive, there is often a trade-off—those with good genetic resources (healthy offspring) may be less reliable long-term partners. Women with high estradiol, the predominant female sex hormone, are regarded as more attractive. But they are more likely to flirt, to have an affair, to be less satisfied with their partner, and to be the target of "mate poaching."[36] Men with high testosterone are more likely to have an affair, divorce, and delay marriage.[37] They are also less likely to respond to infant cries than men with lower testosterone.[38]

It's worth clarifying a few points here. First, these strategies need not be conscious or explicit goals that we consider when we choose a mate. Rather, much of what falls under the rubric of sexual strategies is little more than unconscious biases and preferences that guide our choices. We may find certain facial features attractive in part because (typically unknown to us) they are signals of underlying genetic qualities. And we may convey sexual signals to others without being fully aware that we're doing so. Scientists have confirmed, for example, that around ovulation women tend to wear tighter clothes, seek social settings more than they would otherwise, and purchase more products to enhance their appearance. They typically do so without being aware of these shifts or that the shifts have to do with short-term sexual strategies.[39] Many people would vehemently deny that these strategies shape their behavior, and are surprised when they learn about the unconscious factors at work behind the scenes.

What is perhaps most striking regarding female preferences for male faces is that their preferences are not fixed, but fluctuate across a woman's cycle and according to their own relationship status. That is, women's preferences shift toward the more masculine faces when they are in the follicular phase of their cycle, when conception is most likely. One explanation for this is that hormonal influences shift the brain response to increase the value of facial characteristics that reflect good genes in a potential short-term partner. This shift in preferences toward more-masculine faces also coincides with the frequency of short-term mating and extra-partner affairs. Sexual affairs are 2.5 times more likely when a woman is ovulating.[40] This coincides with other shifts in behavior during this phase of a woman's cycle. For example, women going to a club during this phase wear shorter skirts and show more skin than they do during other parts of their cycle.[41] Women are three times more likely to wear pink or red when they are ovulating.[42] When presented with photographs of women taken across their cycle, independent judges are able to reliably identify when a woman is ovulating based on assessing when the women were trying to look most attractive. Male partners more actively "guard" their female partners during this time.[43]

These shifts in how women evaluate male faces impact more than just mating strategies, suggesting they play a role in social selection as well as sexual selection. In one study, women were given resumes of various male job candidates that included the man's picture (the faces had been independently rated in terms of masculinity). They were asked to assign the candidates to various job positions, which differed in terms of salary, perks, office size, and so on. The assigned positions shifted across the women's cycle. In particular, they assigned more high-status positions to highly masculine faces when they were near ovulation or ovulating.

Reproduction doesn't have to be the goal of these strategies—it may even be consciously avoided. The fact that we've developed technologies to avoid reproduction doesn't mean sexual strategies are rendered inert. Evolutionarily important goals are seldom the proximate goals of our behavior. Rather, evolution shapes the structure of our rewards to make fitness-enhancing behaviors generally rewarding ones. Most of us prefer sweet desserts over celery stalks, but rarely do we justify a second trip to the dessert bar by pointing out the possibility of a future famine. Of course, evolution may have shaped our appetite to indulge in calorically dense foods because "getting while the getting is good" was a sound fitness-enhancing strategy, but it wasn't a conscious goal. Sweet and fatty foods taste good because evolution has shaped our reward systems to align evolutionarily beneficial goals with reward. Just as we can untether sex from reproduction by introducing contraceptive technology, we can introduce the treadmill and the StairMaster to combat fat storage from caloric intake. But just because we can fool evolution by working off that piece of chocolate cake with forty-five minutes at the gym, that doesn't mean our taste for sweet foods no longer has any sway over us. We can have sex for pleasure only, but the sexual strategies of mate choice still shape our sexual behavior and motivation.

And Then What Happened?

Less than ten years after Mailer's article appeared, the seeds of the rebellion he contemplated were taking root in places like Haight-

Ashbury in San Francisco. In 1965, the *San Francisco Examiner* reporter Michael Fallon changed Mailer's hipster to "hippie" to describe the new residents of Haight-Ashbury and their involvement in causes such as the Sexual Freedom League and LEMAR, the movement to legalize marijuana. By 1967, seventy-five thousand young people had flocked to Haight-Ashbury to take part in the Summer of Love. Revolution—wafting upward alongside the plumes of marijuana smoke—was in the air.

What happened next is open to interpretation. Here are a few clues that something happened to cool as an oppositional counterculture. Consider, for example, 1535 Haight Street. In 1966, at this address, the brothers Jay and Ron Thelin opened the Psychedelic Shop, the original drug paraphernalia shop and a main Haight-Ashbury hangout, which also contained radical books, non-Western philosophy books, and other representative works and goods of the alternative lifestyle. Allen Cohen, editor of the *San Francisco Oracle*, a psychedelic, rainbow-hued underground newspaper, worked behind the register, and in 1967 he was arrested there for selling Lenore Kandel's collection of erotic poetry, *The Love Book*, which the California Supreme Court agreed was obscene. (Kandel would later contribute 1 percent of her royalties to the city police retirement fund as a way of thanking them for all the notoriety, which caused sales of the book to skyrocket.)

The Psychedelic Shop closed its doors long ago, and the address is now home to Big Slice Pizza. But what is more telling is that some years ago, the storefront adjacent to 1535 Haight Street became home to a Gap clothing store, one of the most visible symbols of mass consumer culture (it has since been replaced by the surf-culture shop RVCA). If ever there was a perfect application of mass consumption principles to retail clothing, it is the Gap. With around $16 billion in annual revenues and more than three thousand stores worldwide, Gap Inc. includes Banana Republic, Old Navy, and Gap Baby/Kids. What is more surprising than the idea of a Gap at the epicenter of Haight-Ashbury is the fact that in 1994 Allen Ginsberg, the voice of Beat poetry, consented to appear in Gap's "Who Wore Khakis?" series of ads and worked with them on its production. Other ads in the series used the images of dead figures of cool that Gap had appropriated, including Jack Kerouac, Andy Warhol, James Dean, and Miles Davis. Thirty years earlier, Ginsberg had told America to go "fuck yourself with your atomic bomb." Now he was telling America to buy their pants at the Gap.

Ginsberg's appearance in a Gap ad certainly wasn't unique. For both the hipster and the hippie, music was central to the expression of their rebellion, but the Beatles song "Revolution" has since been used in a Nike ad, an advertising trend that goes back at least to 1967 when Buick offered the Doors $75,000 to use "Light My Fire" in an ad for Opel. Perhaps the most incongruent use of 1960s anti-consumerist anthems was the appearance of the Beatles' "All You Need Is Love" in an ad for a Chase Bank credit card, appearing without a hint of irony. Since 1991, the Grammys have handed out an award for Best Alternative Music Album, raising this question: If the National Academy of Recording Arts and Sciences gives it an award, in what sense can it really be alternative? Bands like Rage Against the Machine may sing about taking down the system, but they are comfortable sharing their profits with such establishment labels as Sony Records. Only the punk band the Sex Pistols refused to show up for their induction into the Rock and Roll Hall of Fame, calling the museum a "piss stain," but for most people that gesture and the incorruptible pose it embodies just holds some nostalgic charm. Indeed, the so-called heirs of punk, Green Day (whose drummer goes by the name Tré Cool), are perfectly comfortable standing onstage accepting five Grammys and are one of the best-selling bands of all time, much to the ire of the Sex Pistols' Johnny Rotten. One of Green Day's albums was made into a Broadway show, and the cast sang in the Macy's Thanksgiving Day Parade.

Cool, then, which Mailer defined in terms of its oppositional stance to consumer culture, has become a chief good of consumer culture, something that those who still believe in an oppositional counterculture deride as part of the problem, not the solution. For that reason, the Adbusters founder Kalle Lasn suggests that cool is nothing more today than "a heavily manipulative corporate ethos." It is the thing that companies pay large fees to trend analysts to find so that they can sell it. It is what we seek when we buy cars, clothes, decide what neighborhood to live in, what job to take, where to go on vacation—just another consumable good. The choice that Mailer presented to the hipster—live in mainstream culture or reject it for life in a dangerous netherworld—is one we no longer even recognize. Being cool today doesn't require dropping out, in part because, as we'll argue, in a pluralist consumer culture there's really nowhere to drop out *to*.

Indeed, a basic misinterpretation persists of cool and its relation to consumerism. Far from being antithetical to consumerism, the oppositional consumerism of cool was a major force in driving the transition from hierarchical to pluralist consumer culture—and continues to drive social change in the form of the consumer politics of ethical consumption. So how did cool go from its stance as an oppositional counterculture to its status today as a central element of pluralist consumer culture? This transition will be the subject of the next chapter.

8

DOTCOOL

September 19, 1985, was the last hurrah for rebel cool and the beginning of its end. On that September day, the Commerce, Science and Transportation Committee of the United States Senate opened public hearings on offensive content in records. Music had long been near and dear to the heart of rebel cool. In academia during the 1970s and 1980s, a particularly influential group of British neo-Marxist social theorists, known as the Birmingham school for the university where they founded the Centre for Contemporary Cultural Studies, saw youth music subcultures—from the Teddy Boys to mods, hippies, skinheads, and punks—as expressions of class struggle.[1] Their members' dress, music, rituals, and language were all forms of symbolic resistance, contesting their subordination by the dominant culture. Think of David Crosby's 1970 song about refusing to cut his "freak flag" because to hippies like him, having long hair was an act of symbolic defiance. In fact, the 1969 movie *Easy Rider* ended with the two bikers lying dead by the roadside, shot by a couple of hillbillies because of their long hair. Now the government had the recording industry in its sights.

This curious showdown between the government and the recording industry was sparked by Tipper Gore's decision to purchase Prince's album *Purple Rain* for her then-eleven-year-old daughter, Karenna. Things started to go downhill when they listened to a song called "Darling Nikki," a song that mentions masturbation and Nikki's proclivity to "grind" (the song's lyrics are remarkably tame by today's standards). Outraged, Tipper joined forces with Susan Baker, the wife of Treasury Secretary James Baker, and other prominent Washington women, who became collectively known as the "Washington Wives," to form the Parents' Music Resource Center (PMRC). They proposed a system of music rating labels, requiring lyrics to be printed on album covers, requiring albums bearing warnings to be placed under store counters, forbidding such works to be broadcast, and asking music companies to reassess their contracts with musicians who produced explicit music. They also released a list of the songs they found most objectionable, which became known as the Filthy Fifteen, along with their proposed ratings categories, based on sex, masturbation, violence, the occult, and drug- and alcohol-use content.[2]

With their Washington connections and ability to create media pressure, the PMRC lobbied successfully for the Senate Commerce, Science and Transportation Committee, including the then–Tennessee senator Al Gore, to hold public hearings on the matter, which came to be known as the PMRC hearings. A month earlier, nineteen record companies had agreed to put "Parental Guidance: Explicit Content" stickers on records, but the PMRC wanted the additional concessions. The first witness was Susan Baker, whose testimony included a litany of rape, teen pregnancy, and teenage suicide statistics, suggesting that "the pervasive messages aimed at children which promote and glorify suicide, rape, sadomasochism, and so on" were contributing factors.[3] The hearings also featured testimony from three musicians protesting the PMRC's attempt at censorship. Frank Zappa, who was not targeted by the PMRC but was an outspoken and articulate voice against censorship, spoke passionately against the PMRC. A year later, his new album *Jazz from Hell* would receive an explicit lyrics sticker—despite the fact that it was an instrumental album with no lyrics (possibly the album title or a song title, "G-spot Tornado," was the offender). Perhaps most incongruous of all was the testimony of John Denver, who noted that his song "Rocky Mountain High" had been banned by radio stations in 1972 over fears that it referenced drug use. It's now one of the official songs of the State of Colorado. Then came Dee Snider of Twisted

Sister, appearing in his hair metal persona; his band's song "We're Not Gonna Take It" was one of the Filthy Fifteen. How times change: when Extended Stay America featured the song in a 2012 commercial, the rebellion was against having to make coffee in a hotel bathroom.

While the PMRC's full recommendations were never adopted, a few months after the hearings a compromise was reached: explicit albums would bear generic stickers, sometimes referred to as Tipper stickers, reading "Explicit Lyrics—Parental Advisory." These would be voluntarily placed on albums at the company's discretion, as they still are today. While some stores refused to carry such albums (Walmart still does not sell them in its stores or online), it's unlikely that such stickers had an adverse effect on sales. What the PMRC did accomplish, however, was to create a stark image of hierarchical authority. Ironically, much of the music of the day was itself consciously creating images of hierarchical authority in its lyrics and videos to invoke the rebel instinct in its listeners. In fact, the Twisted Sister video for "We're Not Gonna Take It" featured the actor Mark Metcalf—best known for his role as the ROTC leader Doug Neidermeyer in *Animal House,* a prototypical square—as an authoritarian father who despises rock music. In the video, his son rises up in defiance to become the cartoonish lead singer of Twisted Sister and sing an anthem of youth rebellion. The image of the buttoned-up Washington Wives protesting the very same songs fed directly into the rebel instinct that drove interest in the songs in the first place. A few days before the hearing, none other than Donny Osmond made the point on *Nightline* that a sticker would make an album cool and kids would want it that much more.[4] He even said he might have to add some suggestive lyrics to his own songs to avoid a dreaded G rating that would doom sales.

There's an ironic coda to the PMRC episode. In 2009, when the surviving members of the Grateful Dead reunited for a concert in Washington, D.C., they were joined onstage by Tipper Gore, who accompanied them on drums for the song "Sugar Magnolia." Tipper had been a drummer in an all-girl rock band, the Wildcats, in high school, and had joined the Grateful Dead on previous occasions onstage and given them a White House tour in 1993. The Gores' friend John Warnecke, an early manager for the Grateful Dead, claimed he and Al Gore had regularly smoked marijuana and listened to the Grateful Dead on Warnecke's stereo while they both worked on the paper *The Tennessean*.[5] Whatever one makes of Warnecke's claims, it's more than a little ironic that the Grateful Dead, perhaps the band most associated with drug use and its glorification, was never subject to PMRC scrutiny, nor were any of the bands of their generation.[6] Talkin' 'bout my generation.

The PMRC thus presented rebel cool with an image of hierarchical authority—however diluted—to challenge. This set the stage for a vibrant alternative music in the late 1980s. As late as 1993, musicians were still invoking the PMRC to highlight the rebelliousness of their music. Perhaps most iconic was Rage Against the Machine's appearance at the music festival Lollapalooza, where they appeared naked onstage, mouths covered in black tape, with the letters P M R C across their chests, one letter on each band member. The alternative music scene still saw itself as outside the mainstream, in a fight against censorship and control by authoritarian powers. Seattle in particular was a flash point for this alternative music scene, as were the post-punk grunge bands associated with the independent label Sub Pop Records.

In 1989, Sub Pop signed a local band to a contract by offering an initial advance of $600.[7] The band was Nirvana. Sub Pop released Nirvana's debut album, *Bleach*, later that year. Dissatisfied with Sub Pop's distribution of *Bleach*, Nirvana signed with David Geffen's label, DGC, in 1990 and began production on a second album with the producer Butch Vig, who would become one of the most highly regarded music producers of all time. In late 1991, they released the first single and lead track from the album, "Smells Like Teen Spirit," in an attempt to build some momentum for the album and for the song "Come As You Are," which was anticipated to have the most crossover appeal. The band's ambitions for the album were relatively modest, but soon "Smells Like Teen Spirit" was in heavy rotation at college and rock radio stations despite the worry of many in the music industry that no one could understand what frontman Kurt Cobain was saying. The accompanying video also went into heavy rotation on MTV (which had originally planned on adding subtitles to the video). The song would go on to become Nirvana's biggest hit, reaching number six on the Billboard singles chart the same week the album reached number one. Nirvana won Best New Artist and Best Alternative Video for the song at the 2002 MTV Music Video Awards.

The success of "Smells Like Teen Spirit" is often regarded as a turning point in the relation between mainstream and alternative cultures. The grunge movement, like punk before it, depended strongly on the authenticity of its music, which in turn depended on its unwaveringly countering mainstream sensibilities. "Smells Like Teen Spirit" itself captures the contradictions in the apparent tension between these two cultures. The song seems caught between a strong ethos of authenticity, commitment, and sincerity and a more postmodern sense of irony that makes the pretense of such a stance absurd. Cobain later reflected that "the entire song is made up of contradictory ideas . . . It's just making fun of the thought of having a revolution. But it's a nice thought."[8] Soon after its release, members of the media were quick to call the song the anthem of Generation X and named Cobain the generational spokesman. Writing in *The New York Times*, Simon Reynolds noted that the song "provides a catch-all catharsis that fits in perfectly with the directionless disaffection of the 20something generation."[9] Nirvana would struggle with the song's success, reluctant to play it at shows and referring to it only as "the hit."

A few years later, on April 8, 1994, Kurt Cobain's body was discovered by an electrician in an apartment above Cobain's Seattle home's garage, dead from, apparently, a self-inflicted gunshot wound, with heroin paraphernalia strewn on the floor nearby.[10] The reasons for Cobain's suicide are no doubt complex, involving a family history of depression (and the strong suspicion that he suffered from bipolar disorder), chronic stomach ailments, and a long battle with addiction. But the story line that got the most attention was the explanation of Cobain's death as a tale of co-option.

Think of the psychedelic bus the Merry Pranksters drove on their LSD-fueled escapades: Tom Wolfe made it a symbol of counterculture rebellion in *The Electric Kool-Aid Acid Test*. Years later, Coca-Cola replicated the bus to sell its Fruitopia drinks.[11] In the case of youth subcultures such as punk and grunge, the narrative is that these subcultures start out as legitimate threats to "the System." They are disruptive and cause moral panic. That is, their symbolic resistance exposes some deep internal contradiction in society and threatens its very core. In response, the dominant "hegemonic" culture absorbs the subculture, turning it into a generator of harmless products that people mistake for the real thing.

The idea of selling out is central to the co-option story. The "culture industry" lures rebel musicians and other artists into its mainstream fold with offers of financial success and fame. Nirvana's success was big enough for many of its fans to accuse the band of selling out. In Cobain's case, this narrative continues by suggesting that the inherent contradictions of the struggle to be authentically alternative after finding mainstream success undid him. Perhaps there was no better symbolic expression of this apparent contradiction than Cobain's appearance on the cover of *Rolling Stone* wearing a T-shirt with the slogan "Corporate magazines still suck." *Rolling Stone*, like MTV, was part of the dominant capitalist culture. These contradictions appeared again in *Rolling Stone* later that year when they dubbed 1994 "mainstream alternative's greatest year," with eight alternative albums topping Billboard charts that year, Lisa Loeb's "Stay" topping the singles charts, and indie artists The Offspring outselling major-label heavyweight Pink Floyd.[12] After Cobain's death, musicians with a less serious style eclipsed grunge music. In particular, a spate of post-grunge alternative Britpop groups emerged, such as Oasis, whose album would make the Vatican's list of the ten best rock albums of all time.[13] If there's a stronger sign of the demise of alternative music than making a Vatican top-ten list, we can't think of it.

A view that's especially popular among anti-capitalist critics, such as Naomi Klein and Adbusters' Kalle Lasn, is that rebel cool became contemporary cool through a process of co-option. The story goes like this: Rebel cool was threatening the social order, so capitalists co-opted it by making consuming itself appear rebellious. Now being a rebel actually drove capitalism. This is the "conquest" of Thomas Frank's *Conquest of Cool*. Once rebel cool had been co-opted, contemporary cool emerged as a commodified fake cool.

The result was that the social order—as hierarchical as ever—didn't really change.[14] There's a real tension between this co-option view of cool and Malcolm Gladwell's second law (that cool can't be manufactured). Co-option critics view people as passive dupes who are controlled by looming social forces, and are essentially as passive as brains in a vat, to echo the critics' often-invoked *Matrix* analogy. Cool is manufactured and pumped into the System, providing just enough titillation to keep people plugged into the controlling computer. In contrast, Gladwell offers a radical version of consumer appropriation: consumers transform and re-create the meanings of products, and in so doing erase whatever might have been the "intended" meaning of a product—intended by its makers and their marketing department. Appropriation is the antithesis of co-option. In Gladwell's case, cool products are the result of a complete appropriation, so that manufacturers and brand managers have essentially no control over the meaning of their products.

There's no doubt that appropriation plays a large role in our consumption. People are active consumers (although a long tradition of consumer theory rooted in critical theory says otherwise).[15] We've already seen creative consumption at work in the case of Harley-Davidson: consumers construct social meanings for a product that its makers never anticipated and that diversify over time. A conversation we had with two British brand managers for the well-known cognac company Hennessy illustrates the point.[16] They relayed how surprised they were when Busta Rhymes and P. Diddy's 2001 hip-hop song "Pass the Courvoisier" resulted in a surge of sales and made the United States the largest cognac market in the world. The company's traditional marketing—think stuffy British gentleman's club—had been oblivious to this new market, which had appropriated cognac, now termed "the yak" and mentioned in more than one hundred hip-hop songs.[17] That consumers actively participate in creating the social meaning of products figures prominently in DotCool.

People really do seem to love dark tales of co-option. But for co-option to be real, youth subcultures must first have genuinely, if symbolically, resisted and protested the status quo. Just as people were explaining Cobain's death in terms of the latest subversive subculture to be co-opted, a new generation of sociologists was beginning to question whether the whole "resistant youth subculture" framing of the alternative-vs.-mainstream debate wasn't a big mistake.[18] Were youth subcultures really as political, subversive, and oppositional as the influential Birmingham neo-Marxist school of thought claimed? When researchers actually talked to punks, mods, and rockers in the mid-1990s, they heard a different story. In particular, many people didn't see their lifestyle choices as acts of political resistance, and there didn't seem to be much coherent political ideology within subcultures. Many people also didn't appreciate being labeled one particular type. There were "mockers" (a combination of mods and rockers), hippie punks, and other in-betweeners. The Birmingham school's emphasis on class-based struggle and symbolic political resistance turned out be a caricature. Even worse, the Birmingham school had almost entirely ignored gender because they were so preoccupied with class struggle. Women as well as artists like David Bowie and Roxy Music were ignored because glam rock dealt with gender and sexual identity, which the Birmingham school didn't see as having any political or countercultural significance.[19]

As a result, by the mid-1990s the very idea of youth lifestyles as oppositional subcultures began to fall out of favor. Sociologists heralded the dawning of a "post-subcultural" era. Looming social structures such as the class system no longer determined a person's lifestyle. People were now active consumers who constructed their own sense of identity, not passive dupes. Consumer culture could even be liberating, because it offered a dazzlingly panoply of lifestyles to choose from once the hierarchical and economic barriers to participating in it broke down. Indeed, rather than enforcing class identities, consumer culture gave young people the opportunity to break out of them.[20]

Popular co-option stories treat contemporary society as though nothing much has changed over the last 50 years (or 150 years, for that matter). Co-option requires a dominating social structure to do the co-opting. But by the mid-1990s, rebel cool and the oppositional consumption it drove had helped transform the well-ordered status hierarchy of postwar America into an increasingly pluralistic, fragmented social landscape. For this reason, by the 1990s rebel cool was tilting at windmills. The so-called mainstream was no longer the dominant culture. In fact, the mainstream itself was disappearing. Although consumers and cultural critics alike invoke "mainstream culture" all the time as a dominating force, today it's more of an intellectual conceit than a real feature of postmaterialist societies. The dichotomies that depend on it—mainstream vs. subculture, commercial vs. alternative, and authentic vs. inauthentic—likewise live mainly in our imaginations. (Obviously, we aren't claiming that the demise of cultural status hierarchy meant the end of economic inequality or social injustice.)[21]

Mainstream culture is supposed to represent the consensual values of the majority, but wherever you look for it today, you find fragmentation instead. Consider American religious culture. Forty years ago, more than two-thirds of Americans were Protestant. Now it's fewer than half. Even among Protestants, the last forty years have seen what a Pew Research Center report describes as a "significant internal diversity and fragmentation, encompassing hundreds of different denominations." Forty years ago, 7 percent of Americans described themselves as religiously unaffiliated. Today, it is 20 percent. Among Millennials aged eighteen to twenty-two, it's more than 35 percent.[22]

It's hopeless to try to extract a mainstream, consensual religious culture from this deep fragmentation. Not only is culture fragmented, it is increasingly polarized. That is, the "moderate middle," those people who endorse a mix of liberal and conservative views—who seek common ground—has diminished rapidly over the last two decades. The percentage of Americans who express exclusively conservative or consistently liberal opinions has doubled during that time. The result is that there is virtually no ideological overlap between Republicans and Democrats.[23] And the differences go beyond politics: liberals typically want to live in racially and ethnically diverse urban neighborhoods, while conservatives want to live in the suburbs among people who share their religious faith. The journalist Bill Bishop and the statistician Robert Cushing call this ideological segregation "the big sort."[24]

The fact that we continue to invoke the rebel instinct even though there's no longer a cultural hierarchy is crucial for understanding the shift from rebel cool to DotCool. In particular, the creative energy of rebellion underlies the shift from cool as a signal of opposition to one of unconventionality. And it opened the possibility for invoking the rebel stance ironically. Recall that your rebel instinct triggers when you *feel* that a person, group, or institution is subordinating you (even if no one actually is). There are a few different ways we might feel as though we're being subordinated. We may invoke the rebel stance ironically by invoking it with the awareness that no one is actually subordinating us. Alternatively, we may invoke it by romanticizing or aggrandizing our own situation. There has long been a temptation to romanticize the rebel as a hero, and a hero needs an oppressor. For example, from Hesiod's eighth-century B.C. poetry to Ridley Scott's 2012 science fiction movie, the Prometheus heroic rebel myth has been a particularly pervasive cultural theme. Prometheus defied the gods by giving fire to man, for which he suffers Zeus's eternal punishment.[25] Of course, the Prometheus myth depends on Zeus as an image of institutional tyranny. The image of Prometheus suffering for his actions was particularly popular among Romantic artists, who invoked the image as an analogy for how they felt they suffered for their art. Today, many anti-consumer appeals in particular evoke such images, especially in romanticizing acts of anti-consumption symbolic protests.[26] Some youth lifestyles still occasionally do so as well—the punk lifestyle especially.[27] Never mind that Malcolm McLaren and his then-partner, the fashion designer Vivienne Westwood, formed the Sex Pistols to front the clothes they sold in their boutique, Sex. In 2013, the musician and writer John Roderick raised the hackles of die-hard punk rockers when he suggested that the philosophy punk stood for was "a fundamentally negative one. Punk only tells us what it hates. It has never stood for anything; it stands against things. It is not an intentional indictment; it is a reactionary spasm."[28] By rejecting society, the rebel was free of serious engagement with its problems—any engagement that falls short of overturning the System commits the rebel's cardinal sin: selling out.

It's one thing to be against the System and offer symbolic resistance in the form of culture jamming or an alternative lifestyle. It's quite another to provide a road map to a different system. Echoes of negative rebellion reverberated during the Occupy movements when the anthropologist and activist David Graeber argued that the movement shouldn't issue any demands—doing so would recognize the legitimacy of existing political institutions.[29] The alternative he endorsed was anarchism: a truly free society with no political institutions, based on mutual aid and self-organization, that would lead to a post-violent world, however hazily conceived. In his words, "Unplug the machine and start again."[30]

Beyond the tempting heroic imagery of the rebel, invoking the rebel instinct—even when it means imagining a dominant foe—remains a powerful way to galvanize group identity. Since this promotes in-group cooperation, it's a strategy that's increasingly used to mobilize movements. In fact, this often reaches absurd heights in politics today as groups try to pin the mantle of hierarchical cultural dominance on their opponents while trying to secure the role of heroic rebel for themselves.[31] During her vice-presidential candidates' debate with Joe Biden, for example, Sarah Palin described herself and John McCain as "mavericks" so many times that it became instant self-parody (and "conservative maverick" is an oxymoron). Despite the lack of cultural hierarchy today, political maneuvering to label opponents elitist has likewise become a popular strategy, from conservatives invoking the mainstream liberal media establishment and even the "hipster elite" to hipsters deriding the mainstream.[32]

To truly understand the transition from rebel cool to DotCool, we need to keep in mind that social pluralism, fragmentation, and the proliferation of lifestyles over the last three decades mean that the rebel instinct is now typically triggered in the absence of cultural hierarchy. In particular, the values cool came to signal shifted in response to the new social selection pressures of life in a knowledge economy and its more pluralistic society. As much as critics insist that inauthentic, flaccid substitutes replaced cool and its original rebelliousness, we view the "post-hierarchy" cool that emerged in the transition from rebel cool to DotCool in a more positive light. To see why, let's look into the rise of DotCool as a response to sweeping social changes in the early 1990s.

The Ascent of Knowledge

People often think of the 1960s counterculture as a failure. These accounts end with images of hippies eventually selling out and assimilating into the mainstream, perhaps best exemplified by Jerry Rubin.[33] But this view misses a fundamentally important point: the epicenter of the 1960s counterculture, the San Francisco Bay Area, would become the epicenter of the computer revolution. This was no coincidence. While many in the counterculture were avowedly anti-technology, others saw computer technology as a liberating force. As Stewart Brand, a central figure in this revolution, wrote in 1995, "The real legacy of the sixties generation is the computer revolution."[34] Inspired by the technophile visions of Buckminster Fuller, the media visions of Marshall McLuhan, and 1960s counterculture ideas, a key segment of that generation would embrace what they saw as technology's liberating potential and shape it with countercultural ideas.

Accounts of the computer revolution tend to focus exclusively on technological changes. But this ignores the ideas that drove the changes. Indeed, more than a few computer technologies were the results of LSD-fueled visions—even Steve Jobs said that taking LSD was one of the most important things he ever did.[35] This isn't the place to recount the complex path from counterculture to cyberculture, and it has been done elsewhere in painstaking detail.[36] But it's worth noting that it was John Perry Barlow, a songwriter with the Grateful Dead for more than two decades, who in 1990 applied William Gibson's sci-fi term "cyberspace" to what would become the Internet and who wrote a defining document, "A Declaration of the Independence of Cyberspace," in 1996. The distributed, decentralized structure of the Internet is itself the legacy of a distrust of central authority. Today, organizations such as the OpenNet Initiative monitor Internet surveillance, and Psiphon provides circumvention solutions for users in countries that censor the Internet.

Our everyday computing also relies on a direct legacy of the 1960s counterculture: the Free /Open Source Software (FOSS) that comes from such efforts as the MIT hacker Richard Stallman's GNU project and Linus Torvalds's Linux. Unlike proprietary software, such as Windows, FOSS allows users to modify it and to use it in any way—except that they can't place more-restrictive use conditions on it.[37] Companies such as Oracle, IBM, and Google have embraced FOSS software. For example, Android, the world's most popular mobile operating system—used on more phones than all other operating systems combined—is FOSS. The most popular web server software, Apache, used by more than 100 million websites, is also FOSS. FOSS introduces extremely interesting property rights issues, since it blurs questions of ownership. Indeed, "Is Linux a form of communism?" is a question that's covered in the Linux FAQs portion of their website.[38] These ideas have played a driving role in the democratization of technology, finance, and information, and in democracy itself.[39]

While the computer revolution didn't overthrow capitalism (the measure of success according to some critics), it profoundly changed our world—and it was a major driving force in the transformation from rebel cool to DotCool. DotCool rose in response to the social and economic realities of the 1990s to signal the traits that drove social selection in the new knowledge economy. A seminal event for this transformation occurred in October of 1994: Netscape Communications launched its Navigator web browser. Microsoft would release its web browser a year later. Although the Internet had been in existence for many years, its actual commercial use was prohibited until policy changes were made in 1991 and 1995. Navigator was also the first browser to build in encryption to allow secure e-commerce, setting the stage for the commercial transformation of the Internet.

The spread of the Internet spurred on the rise of a knowledge economy that had been developing at least since the 1970s. One way to trace the rise of the knowledge economy is by examining changes in the number and kinds of patents. Patents that depend on knowledge production first accelerate around 1983 and then sharply after 1994 in knowledge-intensive sectors, particularly the computer, biotech, pharmaceutical, and surgical sectors. At the same time, there is a move away from industrial patents (ones relating to metalworking peak around 1974). But the rise of the Internet and new media involving digital interactivity truly symbolized the transition to the knowledge economy. These technologies fundamentally alter the nature of human communication and interaction, resulting in profound social, political, economic, and cultural changes, and they have played a vital role in the proliferation of lifestyles over the last few decades.

From Rebellious to Unconventional

The rise of our knowledge economy caused a seismic shift in the qualities and skills people need for success today. Whereas natural resources and physical inputs are the backbone of an industrial economy, intellectual assets, human capital, creativity, and innovation are key in a knowledge economy. The growing divide between the 45 million U.S. workers with a college education and the 80 million workers without one reflects these differences. In 1980, those with a college degree earned about 40 percent more than those with high school degrees. Today, it is about 80 percent. One reason for this growing gap is the relatively short supply of knowledge workers—our college graduation rate has not been keeping up with the growth of our knowledge economy.[40] Indeed, as the MIT economist David Autor notes, although most discussions of income inequality focus on the gap between the top 1 percent and the remaining 99 percent, the growth of income inequality *within* the 99 percent, due to differences in educational access and attainment, is equally consequential.[41]

Cool's social signals changed to reflect the qualities highly valued in a knowledge economy. Whereas rebel cool was all about rebelliousness, DotCool is about unconventionality, creativity, and learning. DotCool also became far more ubiquitous than rebel cool, since its norms now align with success in our economy. Let's start to examine the shift in cool's signals by looking at the importance of innovation in a knowledge economy and its link to unconventionality, which people use to signal their capacity for innovation.

As the MIT economist Daron Acemoglu and his colleagues have studied, a company or even an entire country is innovative to the extent that it is open to new ideas and unconventional practices.[42]

Firms and countries that discourage deviations from established norms and impose a set of rigid rules—that is, that are more hierarchical—are simply less innovative. And being innovative is critical for economic success today, as the graveyard of corporations from Blockbuster to Kodak attest (Kodak invented the first digital camera, but didn't develop it for fear of hurting the company's film business). Today's emphasis on innovation is in stark contrast to the 1950s Organization Man, who was portrayed as giving up his individual creativity for the collective outcome. His conformity and unquestioning loyalty were in exchange for the promise of lifelong stability, which sheltered him from risk and uncertainty. He was dying the slow death that Mailer warned of, and helped spark the impetus for rebel cool. Organization Man was, above all, conventional—following the norms that shaped the appropriate behavior, attitudes, and even thinking of his day.

Today's knowledge worker is valued for his unconventionality, because originality drives innovation. This new ideal is reminiscent of the sociologist Paul Halmos's description of positive rebellion as the "creative adventure of the individual beyond the boundaries of the conventional."[43] Being unconventional involves "thinking outside the box" to develop new norms, behaviors—and material goods.[44] It's no wonder, then, that a culture that encourages and endorses unconventionality would emerge alongside a knowledge economy. Indeed, an iconic 1997 ad from Apple captures the cultural shift from rebellion to unconventionality. As the voice of the actor Richard Dreyfuss testifies to the power of "thinking different" to change the world, we see figures of the twentieth century, including Albert Einstein, Bob Dylan, Frank Lloyd Wright, John Lennon, Richard Branson, Amelia Earhart, Muhammad Ali, and Pablo Picasso. The ad was part of Apple's "Think Different" campaign, which itself signifies the prominence of unconventionality. The contrast in tone and imagery between this ad and the company's 1984 dystopian ad is remarkable. Instead of changing the world by violent rebellion, now the imagery focused on changing the world by challenging norms and convention.

Today's emphasis on unconventionality also creates less hierarchical, cooler workplaces, and maybe even cool cities. Interest in creativity and innovation is so intense today that even major neuro-science initiatives have sprung up to try to understand how to foster it.[45] We're a long way from understanding how the brain innovates, in part because it's a complex social process. Innovation depends not just on the isolated creative person but also on her interactions with others and the surrounding structural conditions. Consider the contrast between Vincent van Gogh and Pablo Picasso. Van Gogh created around nine hundred paintings, but sold only one in his life-time. In 1907, Picasso showed a small group of friends his painting *Les Demoiselles d'Avignon*, which many art historians now con-sider the first cubist work and one of the most important works of the twentieth century. His friends reacted with shock. Within five years, however, the art world would recognize Picasso as the founder of cubism, and a revolutionary form of art was launched. Picasso's emergence wasn't due to self-promotion. But Picasso was develop-ing cubism during a time of fragmented art markets, which created many market niches, lowered the costs of experimentation, and made dealers more receptive to radical artwork.[46] Early cubism itself was also fractured, with one group of artists showing only at salons and another at private galleries. Daniel-Henry Kahnweiler was the exclusive dealer for Picasso, which meant the artist was also insu-lated from public reaction, allowing him to take risks with his work.

Timing isn't everything, but studies such as those of Picasso and the emergence of cubism have led to a greater appreciation for the structural conditions that foster radical innovation. We see a similar sort of interest in how our surroundings foster innovation in the rise of unconventional workplaces today. Nowhere is this more striking than in corporate workplaces such as the Googleplex, the corporate headquarters of Google in Mountain View, California, though "cool" work spaces are increasingly common in the tech sector.[47] It's no surprise, then, that a society that increasingly aligns unconvention-ality with economic success would become less hierarchical over time. Initiatives to promote and foster postindustrial creative cities have sprung up all around the world.[48]

The idea is to view urban renewal as a process that creates the city features that foster a creative economy. Just as industrial cities attracted industry by constructing the necessary physical infrastructure, the vitality of cities in a knowledge economy depends on their creative climate. Indeed, the transition from an industrial to a knowledge economy spurred an enormous migration from industrial cities to postindustrial ones, reshaping the economic landscape.[49] Many industrial cities suffer as a result of these changes, while postindustrial cities boom.[50] In fact, although having a college degree means a wage premium, the size of that premium depends on whether you live in an industrial or a postindustrial city. For example, the average salary of someone with a college degree in Stamford, Connecticut, is $133,479. In nearby Waterbury, Connecticut, the average is $54,651. The difference is due to the fact that in Stamford 56 percent of the population has a college degree. In Waterbury, it's 15 percent.[51]

The Kid in the Gray Hoodie

One of the starkest contrasts between 1950s rebel cool and Dot-Cool is how they relate to work. Rebel cool and work were diametrically opposed. Today, policy makers, academics, and workers regard knowledge work not only as creative and cool, but as an avenue for self-development. As Rosalind Gill notes, work at the aesthetic end of new media, such as digital animation, web design, and digital arts and design, is regarded by young knowledge workers as "cool, creative, and egalitarian."[52] One of the deepest tensions in rebel cool—that between making a living and authenticity, since getting a job suggested selling out—has been dramatically defused in a knowledge economy.

People regard knowledge work as cool in part because its products reflect creative thinking. Technology journalists describe new robotic surgical devices, for example, as "cool"; programmers blog about "damn cool algorithms" (who wouldn't think the latest algorithm for homomorphic hashing was cool?); new media designers offer "cool" HTML effects to add to your website; computer industry journalists make frequent lists of "cool" apps; people post "cool" DIY projects on Pinterest; and business writers recommend new start-ups that aren't just cool, but "wicked cool."[53]

By eliminating the tension between work and cool and embracing unconventionality, DotCool has also helped elevate the traditionally derogatory terms "nerd" and "geek" to positive terms.[54] The online retailer ThinkGeek sells more than $100 million of geeky cool toys, clothing, and other paraphernalia every year.[55] In fact, a 2012 online campaign in Sweden urged the Swedish Academy to alter the definition of "nerd" to a more positive one. Mark Zuckerberg insists he's not cool, saying, "I'm like the least cool person there is,"[56] and he's obviously right if the cool in question is rebel cool. One thing Zuckerberg is, though, is unconventional, which has spawned management books touting the power of his type of leadership.[57] Indeed, the ever-present hoodie that Zuckerberg wore while on his IPO tour raised the issue of whether he was *too* unconventional, and the episode became known as Hoodiegate.[58] In addition, with the transition to a knowledge economy and the resulting shift in social norms around unconventionality, there's a growing fusion of cool and geek culture. Consider the rise of "geek chic," often typified by thick-framed glasses and a cardigan. Some might go so far as to suggest that hipsters now appropriate geek styles—and trying to define the differences and similarities between geeks and hipsters has become a blog staple.[59] Actors such as Michael Cera and Ellen Page represent the prototype of the cool geek. Nerdcore hip-hop has long been a musical style, and musicians like King Krule (Archy Marshall) blend geek and cool.

The changing connotations of "nerd" and "geek" reveal the new socially valued traits in a knowledge economy. Physical strength and aggression, for example, are less important in a knowledge economy and may even become liabilities. One of the most dramatic illustrations of this is the emergence of athlete hipster fashion in virtually all professional sports, including baseball (consider the hipster beard of Brian Wilson), football (the trademark nerdy glasses of Von Miller), basketball (the pink pants, printed shirts, and red glasses of Russell Westbrook, the NBA leader of geek chic, though LeBron James and just about everyone else are close behind), and, most incongruous of all, hockey (the ironic mullet of Kris Letang).

If cool's values shifted in a knowledge economy as fundamentally as we've suggested, then we should expect to find that people today really do mean something different when they talk about cool. In fact, a 2012 study revealed that the traits people now associate with cool differ dramatically from those of rebel cool.[60] Ilan Dar-Nimrod and his colleagues asked approximately 350 people to list the attributes they associate with the word "cool" (they could list anything they wanted). When the researchers compiled the responses, they found that people associated cool with friendliness, personal competence, trendiness, desirability, attractiveness, unconventionality, caring, honesty, humorousness, confidence, emotional control, and hedonism. This surprised the researchers, since people didn't seem to associate cool with its traditional associations, such as being rebellious, tough, or detached. When they asked another group of people about these traditional associations, they found that the old traits were no longer socially desirable. Cool certainly didn't appear to mean what it used to. These changes reflect the rise of a new sense of cool, far less rebellious and far more prosocial and creative.

Hipster 2.0

We've seen that cool means something very different now. But what about our claim that cool is more ubiquitous today, since its norms are so valued in our economy? One clue is the fact that, according to a social-media poll of thirty thousand people in fifteen nations, even *countries* can now be ranked in terms of cool, with the United States the coolest and Belgium the least cool.[61] Consider also the omnipresent hipster. A twentysomething man wearing thick-rimmed glasses is filling up his car with gas. He sports a stubbly beard, hair gelled up and back. He's wearing a leather retro jacket, but he looks a bit nerdy. His name is Scott, and he turns to the camera to tell us why he gets ticked off at politicians who put unnecessary regulations on employers, because his friends want jobs, not empty promises. That's why, he tells us, he's a Republican.[62] Scott the hipster Republican is the latest attempt by the GOP to attract Millennial voters. Almost instantly, the ad was widely parodied, including a skewering by John Oliver on his show *Last Week Tonight*.

What's most interesting to us about this ad—and the reaction to it—is the emergence of the hipster as the stereotyped representation of the Millennial Generation, making Scott an important political demographic. The last three decades of cultural fragmentation and diversification make it difficult to generalize about DotCool personalities, and we'll resist invoking stereotypes of the hipster. It's clear, however, that the new images of hipsters are a far cry from Mailer's hipster. The newer images started to emerge in the 1990s, and by 2003 were prominent enough to be satirized by the Williamsburg author Robert Lanham in *The Hipster Handbook*. Cultural interest in the hipster took off around 2008 and now has a global reach.[63] Hipsters can be found across the globe, from Jakarta, Bangkok, and Shanghai to Dubai. In a recent survey, half of the respondents aged eighteen to twenty-nine identified themselves to be hipsters.[64] In the same poll, only 16 percent of Americans had a favorable opinion of hipsters. Since only 5 percent of respondents outside of the eighteen-to-twenty-nine-year-old demographic identified as hipsters, much of the disdain for hipsters appears to be intergenerational. Indeed, the disdain is also international. In Dubai, hipsters are mocked with faux news stories about how they have to be rescued when their skinny jeans tighten up in the desert heat.[65] It's easy to satirize hipsters, just as every era has its stereotype, from hippies to yuppies to slackers. Despite being an easy target, hipsters illustrate many of the qualities that distinguish DotCool from rebel cool, as we examine below.

Work and the New Social Contract

We've established that much of the work in a knowledge economy is viewed as innovative, creative, and even cool. It can also include pervasive insecurity. Whereas Whyte saw the Organization Man as forsaking his individualism for the collective security of the corporation, the situation today is almost flipped on its head. The new social contract between employees and employers has erased the promise of stable employment, as documented powerfully in Andrew Ross's book *Nice Work If You Can Get It*. Daniel Pink's *Free*

Agent Nation presents the new work realities in a sunnier light, emphasizing the positive aspects, which are more applicable to highly skilled work that remains in relatively short supply. The average worker stays in a job for 4.4 years. Almost all Millennials expect to stay in a job for less than three years.[66] These changes place an enormous premium on education as well as on the capacity to keep pace with changing technologies and knowledge flow. The influential knowledge economy sociologist Manuel Castells coined the term "self-programmable labor" to describe workers who can virtually reconfigure themselves autonomously in light of a constantly updating world.[67]

Among the traits that have been elevated in a knowledge economy are those underlying "self-programmable labor," the learning abilities necessary for innovation, creativity, and self-directed learning. A particular suite of cognitive and emotional processes underlies these capacities. One way to think about them is by considering a central distinction that's made in the field of intelligence research: that between "fluid intelligence" and "crystallized intelligence." The psychologist Raymond Cattell introduced this distinction in the early 1970s to measure the capacity to solve problems in new situations (fluid intelligence) and the ability to use acquired knowledge (crystallized intelligence). Many scholars note that Industrial Age work emphasized crystallized intelligence, in which workers learned a set of skills and then applied those throughout their career, a model that is inapplicable to a knowledge economy.[68] One way to measure this is by estimating the half-life of a field, which is the amount of time it takes for half the knowledge of a field to "decay." This used to be twenty years for many fields, but today it is closer to three to five years.[69] Indeed, the major education reform today, known as Common Core, emphasizes critical-thinking, problem-solving, and analytical skills over memorization and crystallized knowledge. Even the creators of the SAT, a test long considered too dependent on memorizing information, have announced a major overhaul to its format, with greater focus on critical-thinking and analytical skills.[70]

Of course, we need not only to "self-program," but also to demonstrate this constant updating to both current and prospective employers or clients. As Rosalind Gill succinctly puts it, "Life is a pitch." These new realities have crucial implications in terms of the social signaling framework we've been exploring, as they predict new kinds of signals relating to knowledge economy skills and an intensification of signaling due to changes in the nature of careers. This is one reason why the language of branding has crept into the description of personal identity. "Be your own brand" is a mantra of the knowledge economy, as it requires negotiating your online presence, packaging and presenting your skill set in a world of uncertain job security, and more.

Given the central importance of these traits in a knowledge economy, how can we signal them?

Inconspicuous Cool

Rebel cool and the cool consumption it spawned involved conspicuous signaling. Indeed, rebel cool depends on making your oppositional consumption transparent to your own in-group as well as to out-groups. Some rebel cool consumption, such as punk style, was downright "in-your-face" consumption that depended on getting a reaction from non-punks. That's the reason why styles tend to get more outrageous over time, as evidenced by the Juggalo lifestyle today. These are fans of the music group Insane Clown Posse, who sport clown makeup, hair resembling spider legs, and widely varying amounts of clothing (Google at your own risk).

Although some researchers view contemporary cool in terms of conspicuous signaling, DotCool differs from rebel cool in its use of inconspicuous signaling.[71] While conspicuous signaling is the most dominant form of consumption, some signals are inconspicuous, meaning that they are intended only for a specific audience. This is also true of certain biological signals animals use. Some warning calls, for example, are intended to be recognized by members of their own species and not by predators. Inconspicuous signals are often used within human communities that fear persecution if they are identified, as was the case with the "hanky code" that originated among gay men in the 1970s. By using a system of colored bandannas either tied around a belt loop or hanging out of the back pocket, they could communicate sexual preferences to other gay men while others around them remained oblivious.

In both cases, inconspicuous signaling stems from the need to disguise signals to avoid detection by outsiders. However, it can stem from other motives beyond fear of detection. The signal itself can be a sort of test to find particular types of people to interact with—namely, people who have special knowledge or skills that enable them to detect the signal. As it turns out, many traditional luxury goods take exactly this form. As the marketing professors Jonah Berger and Morgan Ward found, many of the most expensive luxury-brand goods (such as sunglasses and handbags) have branding that is less, not more, conspicuous. For example, they found that 84 percent of sunglasses costing between $100 and $300 displayed brand logos while only 30 percent of sunglasses priced above $500 did.[72] According to their view, more expensive inconspicuous goods are less visible to a large audience, but are visible to a more select set of insiders who have the cultural capital necessary to recognize the good. While Berger and Ward suggest that the end goal of this signaling is to gain distinction, these signals follow the logic of social selection and are affiliative.

The meaning of the signal, or even the mere presence of a signal, is intentionally designed to elude anyone not belonging to an intended audience. Inconspicuous signals can also involve ambiguous and multiple meanings. One meaning can be for the intended audience while the other—we'll call it a decoy meaning—is designed for outsiders. This has long been a staple of fashion. Consider the Italian fashion brand Diesel, founded by Renzo Rosso. Like other designers who use postmodern themes, Rosso intentionally incorporates contradictory elements and styles. Clothes are also deconstructed by using frayed material, messy and ripped seams, and various stressing techniques. Rosso juxtaposes this sort of irreverent fashion to that of a designer like Ralph Lauren, whose fashion signals a consistent, traditional image.[73] Below, we'll apply this to that much-maligned creature of contemporary consumer culture, the "ironic hipster," whose use of vintage clothing likewise incorporates multiple and ambiguous meanings.

Those luxury sunglasses work as inconspicuous signals because they have a high "information cost" along with their high purchase price. That is, they require both the sender and the receiver to have invested a large amount of time learning what is fashionable. So luxury connoisseur consumption uses high purchase price and high information cost to signal traits (owner wealth and cultural knowledge). Not all forms of luxury connoisseur consumption need be inconspicuous—consider Bayreuth, a Bavarian town that holds an annual festival in a theater Wagner himself designed for the performance of his operas. Getting tickets to Bayreuth to sit through Wagnerian opera certainly signals cultural sophistication. But we haven't met anyone who thinks it's particularly cool. It has high information cost, but it's not inconspicuous.

Contemporary cool goods are ones that use inconspicuous signals of high information cost, not purchase price. Rebel cool goods were costly in that they invoked rejection from the dominant hierarchical powers—long hair and tattoos were enough to make someone unemployable at one time. Now, they're practically required at certain "hipster" workplaces, bars, and clothing stores. As the story goes, getting a job required selling out, cutting the hair, and morphing into the object of rebel cool disdain, the yuppie. With the shift from hierarchy to pluralism in the 1990s, this cost of rebel goods obviously declined, and the cost associated with cool goods shifted to their information cost. This is important, since cool consumption signals knowledge economy skills—innovativeness, creativity, and self-directed learning. As Macklemore showed in his smash-hit song "Thrift Shop," it's cooler being able to make a 99-cent shirt from the thrift shop look good than a new one from Gucci. That puts the premium on the information cost of a good.

Tickets to attend the Bayreuth event celebrating Wagner's Ring Cycle are difficult to obtain, to say the least. Getting them involves submitting an order every year, with success taking around ten consecutive years of applications (missing a year puts you at the back of the line). It's certainly possible to not actually like opera and go just to be seen—though in our opinion sitting through eighteen hours of German opera is itself a costly signal, to say nothing of the burdensome application process and other costs. This is a kind of consumption cost, which is more common than you might suspect. Avant-garde performances, such as Philip Glass's *Einstein on the Beach*—a non-narrative five-hour opera without intermissions—demonstrate that the consumption of the opera itself serves as a signal of investment. Art house movies fit the bill as well.

These are, in fact, similar to some economic analyses of religious rituals. Asking if it pays to pray, Bradley Ruffle and Richard Sosis discovered that the amount of time religious individuals spent engaged in religious rituals predicted their levels of cooperative behavior.[74] Behavioral norms, such as dinner table etiquette, are costly because they require investment in social knowledge. Rituals and cultural performances are costly because they require time to perform or consume. Cultural performances also require the social knowledge to engage in conversation, if, say, you want to discuss how horrible Frank Castorf's production of the Ring was with your fellow opera buffs. These sorts of consumption costs are often indirectly built into consumer goods whose "user manuals" reside in specialized cultural knowledge. As our example from the world of Wagnerian opera is intended to illustrate, according to influential theories such as those of Veblen and Bourdieu, consumer behavior is shaped by social class. That is, the symbolic distinctions of hierarchical society were supposed to be maintained by the consumption of such "highbrow" culture (consuming opera depended on having cultural capital that was available only to the upper class). Consumption today no longer follows such social hierarchies, but is shaped by pluralistic lifestyles.[75]

The challenge with consumer goods that have high information cost but not a high purchase price is that they are easy to adopt—once outsiders detect their inconspicuous signal. This is why some consumer goods or trends are coolest when their signaling remains highly inconspicuous. Contradicting Gladwell's law that goods lose all cool once they start to become conspicuous, the rate of abandonment depends on their cost along with the signaling intensity of one's environment (the demographic and geographical factors that influence competition for social selection: population density, composition, age—typically twenty-five through thirty-four—employment, and so on).

We should expect signaling intensity to increase over time as the percentage of Americans living in urban areas continues to rise. Currently about 80 percent of the U.S. population lives in urban areas, compared to about 65 percent when rebel cool emerged. Indeed, as Enrico Moretti notes, knowledge work drives workers to urban areas that have large knowledge-based labor markets. Many educated workers proactively move to centers of knowledge work as a hedge against employment uncertainties. Cities that already have such markets are increasingly at an advantage over ones that don't, which is why innovation sectors are concentrated in a small number of cities. To further exacerbate the disproportionate flow of knowledge workers to these cities, over the last thirty years it's become increasingly common for people to marry others who are like themselves in terms of education, job type, and salary. The result is that single people seek areas with large numbers of potential partners, which Moretti refers to as marriage markets.[76]

Centers with high signaling intensity should see a high rate of inconspicuous signals and an extremely high information cost of those signals. The recent trend of "normcore" in a fashion-focused center illustrates this idea. As Fiona Duncan recounts in *New York* magazine,[77] while walking around her SoHo neighborhood, she started to encounter teenagers dressed like middle-American tourists—stonewashed jeans, comfortable sneakers, Patagonia windbreakers. In her words, "mall clothes." A few inquiries later, she discovered that these teenagers were in fact on the cutting edge of cool. They were embracing sameness as an anti-fashion fashion statement, taking their fashion cues from the ordinary. In the words of one normcore aficionado, it was an aesthetic that was "exhaustingly plain."[78] The implication is that if a fashion writer living in SoHo didn't know what these kids were up to, then it was a case of inconspicuous signaling par excellence.

But Fiona Duncan let the cat out of the bag the moment the article was published. Google Trends, which measures interest in a term according to how many people search for that term on Google, reveals that interest in normcore shot straight up from zero after that article—albeit almost all of it confined to New York City. In a month, there were 55,000 news results for normcore.[79] Thomas

Frank hoped that normcore would cause "a complete collapse of the imperium of cool,"[80] but of course normcore's appeal depended on being inconspicuous. After all, SoHo kids aren't really emulating the TV fashion of Jerry Seinfeld—the face of high-waisted stone-washed jeans—or Barack Obama, known as President Normcore.[81] The difference between Jerry Seinfeld and a SoHo teenager embracing normcore is that the teenager is aware of the difference.

Cool and Information Cost

The fact that interest in normcore was largely confined to New York City illustrates that inconspicuous goods in an environment of dense, high-speed information networks tend not to spread very far before being detected, further fragmenting cool that relies on inconspicuous signals. That said, within a niche, the coolness of a good will often depend on its information cost. This is one reason why, despite the complaints about homogenization of consumer culture under globalization, product segments are flooded with choice. Consider beer. There are thirteen thousand different beers for sale in the United States. In the 1980s, there were four dozen breweries. Today, there are more than twenty-four hundred, owing to the explosion of interest in microbrewed beer.[82] As Barry Schwartz, author of *The Paradox of Choice*, documented, the astounding array of consumer choices available to us causes anxiety and might even make us more dissatisfied, because it increases the information cost of choice.[83] But by increasing the information cost, it also provides opportunity for signaling knowledge economy traits. It's likely no accident that the most choice lies in alcohol products,[84] since they are often consumed in public settings (the Internet is replete with advice on impressing your beer friends, from cheat sheets on types of beer to recommended beer and cheese pairings).

Our capacities for cultural learning, imitation, and conformity are cornerstones of human social life, and evolution researchers regard them as hugely significant innovations in human evolution, which underlie our extreme sociality. Innovation among apes is extremely slow—a new skill, such as using a stick to dip into a termite hill, spreads through an ape colony gradually, since the art of imitation among apes is limited. In contrast, a new idea or skill can spread rapidly among humans, since we are keen imitators and are able to fathom the goals behind others' actions.[85] In this imitation game, some people will gain esteem via their superior knowledge, and others will follow the lead of innovators—they can save themselves the cost of learning some domain from scratch by relying on others who are esteemed for that expertise. There's no doubt that the Internet has radically altered this process, in part by decreasing the cost of information. In her unintentionally normcore-killing article, Duncan notes how the Internet has changed fashion cycles by providing us with new ways to find trends, such as reverse online image searches, and community sites such as Polyvore that promise to democratize fashion.

Gender, Sex, and DotCool

Of all the differences between rebel cool and DotCool, none are as striking as those surrounding cool and gender. Although Helen Gurley Brown's hugely influential 1962 *Sex and the Single Girl* presented advice to the single woman, advocating financial independence and sex before and outside of marriage, female sexual identity remained uniquely conflicted. This was in no small part because both traditional sexual politics and male rebel cool's embrace of hypermasculinity presented problematic notions of female sexual agency. Also, while the 1960s counterculture focused on civil rights and the war in Vietnam, the counterculture movement reflected traditional gender roles. Indeed, for many women participating in the 1960s counterculture, the rigid gender roles they experienced in it sparked their revolt against their subordination in those roles more generally. For example, in 1965 Casey Hayden and Mary King circulated their paper "Sex and Caste" to women active in civil rights and the antiwar movement. It was based on their experiences in the Student Nonviolent Coordinating Committee (SNCC), where "hierarchical structures of power" excluded women from participation in leadership roles.[86] In 1966, Betty Friedan and others founded the National Organization for Women, which helped launch the second wave of feminism's string of legal victories for women's rights.

Second-wave feminists also raised and debated issues of sexual identity, feminist sexual theory, and sexual freedom.[87] By the early 1980s, these debates had transformed into what would become known as the Sex Wars, which pitted radical feminists such as Andrea Dworkin and Catharine MacKinnon against pro-sex feminists such as Ellen Willis and Gayle Rubin. Disagreements involving topics from pornography, sadomasochism, and prostitution to the nature of feminist sexuality and heterosexual sexual expression increasingly polarized second-wave feminism, leading to the emergence of third-wave feminism by the early 1990s. In popular culture, the contested nature of female sexuality was nowhere more polarizing than in the images created by Madonna.[88] Among all the cultural icons of the last three decades, perhaps no one has sparked more debate than she has. Madonna's cultural relevance is intriguing for her complex influences on sex and gender. In November 1984, Madonna's album *Like a Virgin* sealed her global recognition, which had already surged with her performance two months earlier at the first MTV Music Awards, one of the most iconic moments of the 1980s. Madonna would go on to become the bestselling solo artist of all time and would appear on *Time*'s 2010 list of "The 25 Most Powerful Women of the Past Century."[89] Coincidentally, she also topped *Forbes*'s list of the highest-paid musicians of 2013, earning $125 million.[90]

So while Madonna's overt sexuality was criticized by many as antifeminist, the third-wave feminists who emerged in the 1990s embraced Madonna as a symbol of female sexual agency who defined her sexual identity on her own terms and contested the social norm of passive female sexuality, in part by making claim to the sexual prerogatives of rebel cool men. Even as early as her 1984 "Like a Virgin" performance, Madonna parodied the social commodification and fetishizing of female virginity, an issue that would explode a decade later with the rise of abstinence movements, virginity pledges, and Purity Balls, and would continue to be contested by third-wave feminists such as Jessica Valenti.[91] By the time these movements emerged in the 1990s, overwhelmingly from conservative religious groups, they cloaked themselves in a counterrevolutionary rhetoric opposing what they claimed was the dominant culture of casual sexual promiscuity. Some abstinence movements, such as Silver Ring Thing, even try to make chastity cool by "empowering students to resist conformity and choose to be different."[92]

This oppositional rhetoric grew in the early 2000s to include increased scrutiny of the so-called hookup culture of college campuses and the purported emotional destruction such cultures held for women, in particular, who participated in them. Indeed, pseudoscience also regularly entered the debate, such as when in 2009 the sex therapist Ian Kerner wrote an article titled "Can You (and Should You) Have Sex Like a Man?" Writing on *Today*'s health blog, he warned that the female orgasm releases oxytocin, a hormone that predisposes women to attachment, and when attachment is not forthcoming, "orgasm becomes a regretful reminder of the hollowness of the sex that preceded it."[93] Suggesting an even worse picture of the dire emotional consequences of casual sex, the University of Southern California professor Ruth White adds that while a woman's brain releases oxytocin, men's brains release testosterone, which "drives them off to go find some other women with whom to spread their biological material."[94]

These ideas ignore the pluralism of human sexual strategies. Even worse, they simply get the biological facts wrong. They leave out the fact that male orgasm likewise releases oxytocin[95]—indeed, oxytocin appears to play a central role in male monogamy.[96] In fact, the only longitudinal study we are aware of that examines the effects of hookups on college student well-being found that women and men reported identical rates of casual sex, and women were not more motivated by the thought that hookups might result in long-term relationships, indicating casual sex has equal appeal for both sexes. They found no negative effects on well-being, which is consistent with similar research on adolescents and young adults. The single negative link between hooking up and anxiety they found was among men, which they suggest may be related to male performance anxiety.[97]

As the influential professor of sociology and gender studies Michael Kimmel notes, the last forty years have seen "the most dramatic and most rapid transformation of gender relations in our nation's history."[98] This transformation is reflected in pervasive structural changes in social arrangements, including the workplace. The percentage of women in the workplace is about double what it was in the 1950s, according to the U.S. Department of Labor. Approximately 54 percent of women over the age of sixteen are employed (and about 64 percent of men).[99] Women have now surpassed men in college graduation rates, and about 37 percent of women in the workforce have a college degree, compared with about 35 percent of men. Since women are twice as likely as men to work part-time, men with college degrees outnumber women among full-time workers, although this is expected to reverse in the near future. The increasingly mixed-gender workplace has played a role in changing social norms as well, although the tech sector lags behind.[100] As Kimmel also notes, there has been a pronounced increase in the rise of cross-sex friendships, especially among Millennials. This is particularly interesting, as it suggests that social selection and not sexual selection is playing a larger role in cross-sex dynamics.

One of the most striking differences between rebel cool and DotCool is the extreme shift in the gender composition of the hipster population. Mailer's hipsters were almost all male and invoked a hypermasculine sensibility, resulting in a homosocial way of life, including the mostly all-male adventures on the road. As Lauraine Leblanc chronicles in her study of female participation in the punk culture of the 1980s and 1990s, punk culture was also highly male-dominated and valorized the norms of adolescent masculinity. The same bias was reflected in the literature: the few existing accounts of women's participation in these cultures were asides in studies by male ethnographers who were almost entirely focused on male participation.[101] Today, women are more likely to consider themselves hipsters than men by a large margin, 16 percent to 4 percent. Women also have a more favorable impression of hipsters than men, 21 percent to 11 percent.[102] As in the studies of Harley-Davidson microcultures, the shift in gender relations within these cultures is perhaps their most salient commonality. We suspect the shift to more prosocial associations with cool today reflects these changing gender relations.

Act Your Age?

Jack Weinberg's remark during the 1960s Berkeley Free Speech Movement, "Don't trust anyone over thirty," succinctly captured the sentiment that rebel cool was for the young. Today, however, many of those who make it onto lists of cool people are over thirty, such as fifty-plus-year-olds Brad Pitt, Madonna, Johnny Depp, Tom Cruise, George Clooney, and musicians well into their forties, including Dr. Dre, Gwen Stefani, Ice Cube, Pharrell Williams, and Sean Combs. A blog called *The Hipster Mom*, with the slogan "Have Kids. Keep Your Cool,"[103] comes in from Williamsburg, Brooklyn, ground zero for hipsterdom. There are even hipster-parent baby-name books, such as *Hello, My Name Is Pabst: Baby Names for Nonconformist, Indie, Geeky, DIY, Hipster, and Alterna-Parents of Every Kind*.

All of this focus on middle-aged cool people has led some to ask why we don't just act our age. The crankiest of these is the political theorist Benjamin Barber.[104] According to Barber, consumerism has created an ethos of infantilization, with an aim to suspend consumers in a state of perpetual childhood. Marketers and merchandisers are "hoping to rekindle in grown-ups the tastes and habits of children so that they can sell the useless cornucopia of games and gadgets for which there is no discernible 'need market' other than the one created by capitalism's frantic imperative to sell."[105] Among the apparent signs of impending social collapse, Barber points to businessmen in baseball caps, policemen giving out lollipops to travelers at airports, adults reading Harry Potter books and playing video games, adults watching *Shrek*, and people having sex without reproducing—all signifying what he describes as "puerility without pleasure and indolence without innocence." Barber also contends that college degrees are simple to come by these days (just get one online), athletic success is now often just about taking steroids and "showboating," and even the consumer market is designed to make choices easier than ever before.

The problem is, not a single charge Barber makes is true. Consider, for example, his complaint about video games. This reflects nothing more than a generational prejudice. In fact, when the cognitive neuroscientist Daphne Bavelier started looking at the effects of first-person action video games on the brain, she found that they improved perception, attention, cognition, vision, and multitasking.[106] In contrast with many other forms of learning, these improvements transferred beyond video games to other tasks. Video games do not mean we are simply idling—they are among the most powerful and complex learning environments ever created. They are the template for new technologies that promise to unlock brain plasticity in the elderly, and may be tomorrow's technology of choice for cognitive rehabilitation and offsetting dementia.

Regarding athletic success, Barber's view is patently false. The increase in salaries in almost every sport over the last three decades has intensified competition, while the internationalization of sports has expanded the pool of athletes to the point that extremely specific anatomical specializations determine athletic success. Performance-enhancing drugs in sports only make winning easier when they are used by a small percentage of participants. As the Lance Armstrong saga reveals, performance-enhancing drugs typically spread quickly throughout the entire profession (as increases in the average speeds of races reveal). Athletes take them not so much to gain a competitive advantage but because their competitors might be taking them. This is a classic social dilemma akin to the Prisoner's Dilemma we examined in chapter 4.

Barber's assertions about consumer markets fare no better. As we've examined, consumer markets have become increasingly complex: just note the explosion of boutique beer companies, fine wines, gourmet food, and other goods. In a TED talk, Barry Schwartz relays an entertaining account of his trip to buy a new pair of jeans.[107] He notes that in the old days you went into a store and chose from a few different pairs of Levi's. Today, a single store has dozens of brands, various fabrics, low waist, high waist, skinny leg, relaxed leg, boot cut, and so on. The fine variation in choices significantly complicates consumer decision making. In a classic study, researchers set up two displays of jams at a supermarket. One display featured twenty-four different jams. The other featured six jams. More people stopped to sample jams at the twenty-four-jam display, but only 3 percent of them actually purchased one of the jams. In contrast, 31 percent of the people who sampled one of the six jams ended up purchasing one.[108]

We've also become more sophisticated and have far more information and knowledge about product categories than ever before. Consider the process of buying a car. Up until 2006, car buyers visited an average of 4.1 dealerships before purchasing a car. Today, the average is 1.3. That means consumers know what they want before they even step on a lot. They also have more information than ever before about the invoice price, and they can dictate prices. The result is that profit margins for dealers have dropped, and the proverbial car salesman is fast becoming obsolete.[109] In terms of work, knowledge economy work requires more skills and discipline, not less. Far from lying on the couch all day playing video games free of responsibility, people work more today than ever before. Indeed, a problem Barber and like-minded critics fail to address is that all this consumption couldn't be taking place if people weren't being productive.

This view is what we've called a declinist view. Barber's narrative and others like it follow the typical logic of imagining a better time, lamenting its loss, and predicting our inevitable collapse. But there's a much simpler explanation for the apparent rise in childlike and adolescent behaviors. The skills necessary for knowledge work are all associated with childhood and youth. So it makes sense that workers in a knowledge economy would embrace these traditionally childlike activities to signal those skills. Indeed, these associations are one reason why knowledge-based companies, especially in the tech sector, incorporate so many elements of play into their workplaces, from Google's gingerbread men and donut statues to their workers' extensive collections of toys in their workplaces.

In fact, the growing associations between childhood activities and creativity and innovation reflect one of the major changes in neuroscience over the last few decades. Twenty-five years ago, neuroscientists didn't believe that the brain could change its structure very much past some early critical periods in infancy. While this theory reinforced the idea that childhood was a special period of learning, we now know that the brain's capacity for change may be lifelong. This was the focus of Steve's original research and the origin of "cultural biology."[110] Although mature forms of plasticity are not as pervasive as childhood forms, the idea that the brain is plastic has had enormous social impact and has spawned companies like

Lumosity, which claim to design games that trigger plasticity. While we won't enter the debate over whether these claims are substantiated, a growing awareness of plasticity has stressed the continuity between the child's brain and the adult's brain. This is one reason why play and games have become increasingly regarded as important forms of brain stimulation. Indeed, play, creativity, and health are growing areas of interest in neuroscience, especially in light of an aging population.

Ironic Cool

In 2012, Christy Wampole, a professor of French and Italian at Princeton, sparked a firestorm of debate with her article in *The New York Times* titled "How to Live Without Irony."[111] Claiming irony the ethos of our age and the hipster the archetype of ironic living, Wampole chides the ironic Millennial Generation and offers them tips on learning how to live without irony. But in our opinion, Wampole's beef isn't really with irony. Rather, Wampole is really just a younger version of Benjamin Barber. Her screed is another declinist narrative, which, like Barber's, sees infantilizing as the root problem. In fact, she says as much in an interview about her article when she describes the United States as a self-infantilizing population.[112] It's not irony that's at the root of this self-infantilization. Instead, she diagnoses it as the need to treat everything as a joke. She describes the Millennial Generation as only knowing how to engage through mockery and scoffing. While Barber looks back to a distant time, Wampole looks back to her own halcyon formative era, the 1990s, when people took things seriously, when the grunge movement was earnest, and when young people were hopeful. Leaving aside issues of historical accuracy (David Foster Wallace declared irony the ethos and problem of his age in 1993), we think Wampole is best understood as a reiterator of the eternal cranky intergenerational refrain: "Kids today, why can't they grow up?" Actual data on the Millennials whom Wampole so summarily dismisses as self-infantilizing shows that they are harder workers, better community builders, more service-oriented, more committed to social and environmental justice, and more embracing of diversity than Wampole's Generation X.[113]

Let's turn now to the issue of irony and ironic consumption. As Alanis Morissette inadvertently demonstrated in her 1995 hit song "Ironic," much, if not most, of what passes as irony isn't ironic at all. Likewise, much of what Wampole complains about—in her words, mockery and scoffing—aren't necessarily ironic either. The same for snark, the blend of sarcasm and cynicism that's so prevalent on cable news and talk radio. So what is ironic? Despite what Morissette sang, a black fly in your chardonnay isn't ironic. Rain on your wedding day isn't either. They're just bad luck. Now, if the wine was purchased specifically to repel black flies, or if you carefully chose a wedding location where it never rains, then yes, it's ironic, as a corrected version of Morissette's song illustrates.[114] What makes it ironic in those cases is the fact that the outcome is contrary to what was intended or expected.

As these examples of situational irony suggest, irony has to do with discrepancies between intention and outcome, or, in the case of verbal irony, between what one says and what one actually means. There's another word for a discrepancy between what you say and what you mean: "misunderstanding." So what's the difference between irony and misunderstanding? Two main things: In the case of misunderstanding, an audience only comprehends the surface meaning of what you said. Your intended meaning escapes them. In the case of irony, they comprehend both meanings. But to make it ironic, they also have to recognize that you intentionally said something that had multiple meanings and that this intentional ambiguity had some effect on the discourse.

Irony, then, involves the intentional use of multiple meanings. Indeed, an ironist, such as Jonathan Swift, intentionally uses ambiguity to be partially misunderstood to create a dynamic interaction between the meanings (dialectical movement), a form of indirect communication that is richer and more revealing than direct communication. What often gets lost in the discussion of irony is that particularly effective irony is often misunderstood by some of its audience, who take its literal meaning as its intended meaning. For example, Stephen Colbert had to invoke Jonathan Swift after he was accused of racism for a tweet mocking Dan Snyder of the Washington Redskins. Kazakhstan issued official statements decrying the character Borat, not realizing that Sacha Baron Cohen's satire was directed at the United States, not at them.

So ironic consumption has to do with infusing products with double meanings. We mentioned consumer appropriation earlier in this chapter as involving the consumer's altering the intended meaning of a product. Through how the product is used, it may come to have new meanings, and the original meaning is no longer attached to it. For example, punk fashion appropriated safety pins to use as jewelry. In ironic appropriation, the product retains its original meaning, but comes to gain new meanings or connotations as well. To be ironic, these meanings must contradict each other. Some ironic appropriation is likely to be a form of inconspicuous signaling. In this case, some members within a group introduce a new good and its ironic meaning diffuses through the group, at which point it may lose its ironic cachet and be abandoned. Inconspicuous ironic signaling may be playful, creative, and clever since it deals with plays of meaning.

We suspect that ironic appropriation has become more inconspicuous over time. A decade ago, ironic appropriation was often conspicuous. Goods to appropriate could come from salient outgroups that are positioned as mainstream or lowbrow. For example, a T-shirt or trucker hat from a company such as John Deere could easily be appropriated as a form of negative consumption. This doesn't necessarily mean that there's antipathy for the out-group, only that it creates humorous incongruous meanings (juxtaposing rural and urban lifestyles). When these goods come from dissociative out-groups, like a hipster wearing a Mitt Romney campaign shirt, the ironic appropriation is humorous within the group, but it stems from a negative judgment about the out-group. Culture and fashion critics have been complaining about trucker hats on hipsters since at least 2003.[115]

By asserting that irony was the ethos of our age, Wampole also sparked a backlash from commentators such as Jonathan Fitzgerald, who asserted that sincerity was the ethos of our age.[116] In an age of fracture, the notion that there's a specific ethos strikes us as a nonstarter. And the prominence of irony is overstated as well as misleading. Suggestions that irony somehow squeezes out sincerity with its cynicism are misleading (as T. S. Eliot noted of Swift, "Real irony is an expression of suffering"). Beyond that dispute, there's no doubt that sincerity remains firmly in place today. Indeed, one of the most influential TV series of late, *Lost*, was most striking for its absence of irony and its earnestness in presenting themes that would have been the subject of ironic treatment a decade before.

We think that hipsters are maligned especially by conservatives, who have a higher disdain for them than do liberals—not because of their irony, but because of their earnestness, particularly with respect to issues of ethical consumption. Hipsters are often portrayed as vegans, recyclers, bicycle commuters, "localtarians" (shoppers at local and farmers' markets), obsessed with fair-trade goods. And that leads us to the question of whether cool consumption can be ethical consumption.

Can Cool Consumption Be Ethical?

Throughout this book, we've argued that critics of consumption make many poor policy recommendations because of their mistaken assumptions about and mischaracterizations of consumerism. For example, critics such as Benjamin Barber and Naomi Klein argue that thinking of ourselves as consumers makes us feel and act less like politically engaged citizens. In other words, consumerism "crowds out" our citizenship, making consumerism dangerous to democracy. We should be suspicious of this claim for two obvious reasons. First, political consumerism emerged to play an important role in the creation of American democracy.[117] As the historian T. H. Breen documents, popular images of eighteenth-century colonists as embodying a self-reliant, homespun way of life are myths. Colonists enthusiastically participated in a global economy and had a large appetite for British-made consumer goods ranging from textiles and china to ivory combs and decorated snuffboxes. Indeed, by 1770 the colonies accounted for roughly a quarter of all British exports. When Britain began imposing special taxes on these goods, beginning in 1765 with the Stamp Act, colonists transformed private consumption into political action in the form of consumer boycotts. According to Breen, the colonist exercise of consumer choice in the form of the consumer boycott helped create a national sense of identity that had eluded the otherwise diverse and fractious colonies and provided the crucial link between everyday life and political revolution. Second, many of the most robust democracies today, such as those in Sweden and Denmark, also have the highest levels of political consumerism.[118] The fact is, engaging as political consumers leads to people taking more political action and commitment.[119]

The most egregious and pernicious misunderstanding of consumption is the idea that it's morally suspicious because it appeals to status. Let's take a look at the idea that status seeking is a suspect or even corrupt motivation that should be treated as a pathology. This comes in many forms. One is the Veblenesque idea that status seeking is simply a self-regarding, irrational impulse and an end in itself. Consumerism's critics often claim that these irrational impulses come from "extrinsic" values. These are morally suspect and include a desire to be popular, to get a good image, to have financial success, and to conform. The reason critics call them extrinsic is because they are supposedly "other-directed," in David Riesman's term— directed at impressing others and provoking their envy to fill a narcissistic void, to substitute for healthy self-esteem. Unsurprisingly, these critics claim that consumption feeds, and feeds on, these values, making it complicit in perpetuating pathology. They continue that the better part of our nature comes from our "intrinsic values," which include our desire to affiliate, to accept ourselves, and to enjoy community. Worse yet, these critics charge that extrinsic and intrinsic values are antagonistic, so appealing to an extrinsic value, such as status, harms intrinsic ones.

This line of thinking leads to a critical misdiagnosis of the role of status norms in ethical consumption. For example, the journalist George Monbiot says that activists have tried to motivate consumers to buy green by appealing to their worst instincts—telling them that "by buying a hybrid car, you can impress your friends and enhance your social status." This tactic is a bad idea because it "also strengthens extrinsic values, making future campaigns even less likely to succeed. Green consumerism has been a catastrophic mistake," Monbiot concludes.[120] On the contrary, green consumption, as we will see below, actually turns Veblenesque wasteful signaling on its head. That is, green signaling, akin to competitive altruism, can induce an arms race among consumers to display austerity that minimizes the environmental impact of consumption, which produces social benefits rather than waste.[121]

Monbiot's view, which was also central to a major 2010 report by Common Cause, fundamentally misinterprets the nature of status.[122] Indeed, the deep irony of regarding status as morally suspect is that our emerging understanding of human morality rests on the same theory of signaling and social selection that explains consumption. That is, many moral behaviors signal our commitment to moral norms, such as equity and fairness, for which we gain esteem as moral actors—from which esteem, in turn, we gain high-quality social partners. None of this suggests that we are somehow morally duplicitous. Nor does it suggest that our morality is some sort of disguised scheming, any more than understanding the biological basis of parenting behavior invalidates the genuine feelings of love that parents have for their children.

To see why, consider that there are two keys to our capacity to build and live in complex societies. The first is the extent to which our biology prepares us to internalize and act on the basis of social norms and rewards us right within our own brains for doing so. The second is our collective ability to change norms, some in response to changing environmental and economic realities. Others we change through a process of moral reflection, debate, disagreement, and negotiation, as when we extend notions of justice to previously excluded persons and groups.

What about the intuition that appealing to status results in selfish behaviors that harm society rather than help it—like Monbiot's calling green consumerism a catastrophic mistake? An enormous amount of research debunks this intuition. Consider a few examples. When Switzerland adopted a mail-in ballot system for elections, they discovered that it did not raise voter participation, despite their strong intuition that making voting more convenient and less time-consuming would do so. In fact, in some communities it actually reduced voter turnout. By turning a public behavior into a private one, the mail-in ballot eliminated the esteem incentive of being seen taking time to vote, a costly signal of one's civic engagement.

Similarly, many charities these days raise a large percentage of their revenues by holding athletic events or challenges (runs, triathlons, and so on) for which people pay substantial entry fees or raise sponsorship money. Wouldn't it be easier for people to save themselves the time it takes to prepare and participate in these events and just send a donation instead? Most people who participate in these charity events are not regular exercisers, as indicated by the high injury rates at such events.[123] People participate eagerly in these events because they are costly signals of generosity and altruistic giving. And, a related point, why is it that anonymous donations to charities—despite having the same tax-deduction benefits—account for less than 1 percent of total charitable contributions?[124] As these cases illustrate, esteem incentives are aligned with social benefits. There is no basis for the claim that esteem incentives must be antagonistic to social benefits.

Green Goods Are Identity Goods

Let's examine some economic studies to see how consumption norms and social benefits can be aligned.[125] Economists have increasingly examined the popularity of various models of cars, since they are major household expenditures that also possess strong signaling properties. One of the most studied cars is the Toyota Prius, the most popular hybrid car on the U.S. market, accounting for about half of all hybrid cars on the road. As many researchers have suspected, the popularity of this model may not be due solely to its functional properties or its cost savings, since its purchase price is several thousand dollars more than many standard-fuel cars, and the difference is typically not made up in savings from increased efficiency. In fact, this latter fact may hold a clue to its popularity. The fact that its cost to own is higher than for many other standard cars suggests that it might be a costly signal. It may signal that its owners have sacrificed financially to drive a car that has less of an impact on the environment than a cheaper alternative. There are other standard-fuel cars with more features, such as better acceleration and handling, at the same price point as a Prius, so what accounts for its popularity?

The fact that the Prius is not simply a hybrid version of an existing car model, such as the Honda Civic or Toyota's own Camry, suggests that the Prius is a potent signal because it is so easily recognized as a hybrid car. Indeed, research has found that the Prius is a costly signal in just these ways. But what's crucial about these findings is that the Prius isn't signaling wasteful spending. It is signaling prosocial traits, including altruism and environmental concern. The result is what Steven and Alison Sexton refer to as "conspicuous conservation." The Prius confers social esteem upon its owners, but none of this means its owners are merely using the car for its esteem value. There is nothing in the theory of social signaling that suggests signals are only used strategically in this way. This is why the intrinsic vs. extrinsic distinction is a false dichotomy. We internalize social norms and come to genuinely value them. But this also motivates us to signal these values to others. The possibility that people can fake signals doesn't mean there are no genuine signals. Indeed, the whole point of making them costly is to weed out the genuine from the fake.

One of the most interesting results regarding the Prius as a prosocial signal is that the hybrid cars are identity goods, more than status goods—their signaling value increases as more people use them. By using zip code data, Sexton and Sexton could examine where various car models were more popular. They used voter registration to identify green locales (since Democrats are more in favor of green measures). They found that hybrid cars were more popular in green locales.[126] That's not very surprising, but they also found that the greener the neighborhood, the more Priuses there were. Other hybrids, such as the Honda Civic, were actually less popular in greener neighborhoods. The greener the neighborhood, the more valuable the Prius was—but not other models of hybrids. By using some economic models, the Sextons determined how much of a premium people were willing to pay just to signal that they were green. In some places, the green signal of the Prius was worth $7,000 to its owners.

This explains why places like Berkeley and Boulder, Colorado, are crawling with Priuses. Because the Prius increases in value as more people in a neighborhood drive one, it becomes an identity good. Indeed, it's also one reason why San Francisco has dispropor-tionately high numbers of solar panels, and why some homeowners demand that solar panels be installed so that they are visible from the street even if their house is not in an optimal location for them. Some also prominently display energy-efficient heating and cooling technologies rather than tuck them away in a utility closet.[127]

Linking Consumption and Morality

A central element of our cultural biology view is the role norms play in enabling the flexibility of human behavior. It's likely that linking behaviors to cultural norms was a key to our emergence as a species, as it enabled our ancestors to rapidly change behaviors and social arrangements in response to new conditions. But the reason why this choreographed flexibility works at all is because adhering to norms is socially rewarding and follows the logic of social selection. This is why it's crucial to see that our consumption is linked to norms. Because of this link, consumption itself is highly flexible. Indeed, consider the sweeping changes in consumption since 1950 that we've chronicled in this book. These were all made possible by innovation in consumption norms in response to changing economic and social conditions.

Critics of consumerism contend that consumption's norms are antagonistic to social benefits. For example, they point to norms of wastefulness. But we've already seen that when consumption is linked to conservation norms, people make choices that favor social benefits. Following that norm both supplies status and belonging, and produces social benefits. Of course, the prototypical example of following norms that produce social benefits is our moral behavior. It's not altogether surprising that in recent years researchers have become more aware of the links between consumption and morality. Ethical consumption now includes such considerations as fair trade, sustainability, environmental impact, and consumer citizenship. Ethical consumption illustrates why conspicuous consumption, the hierarchical display of wealth, and wasteful consumption aren't necessary elements of consumerism. What goes unnoticed, due to misunderstandings and biases about consumption, is the *flexibility of consumerism*, its ability to quickly elevate new social norms to become the central determinants of esteem, including moral esteem.

As we saw in the example of the Prius and conspicuous conservation, consumption norms can incorporate strong prosocial signals. It is obviously an open question whether changing consumption norms can have a significant environmental impact. Recent studies suggest that changing consumption patterns could have a significant impact on carbon emissions, in both the short term and the long term.[128] We suspect that altering consumption behavior by changing consumption norms—and potentially eliciting competitive altruism—will be a more effective strategy than admonishing people to simply consume less.

While linking more-sustainable consumption norms to status is a potentially powerful way to change consumption behavior, we also think such norms will have to make a stronger link between consumption and production. Although some forms of ethical consumption, such as fair trade, do make issues surrounding production more visible, most of our consumption remains detached from its production. The problem with this is that our current model of consumption is a linear one, often described as "take, make, dispose," that depletes resources and results in waste at every step. An alternative model, often referred to as the circular economy, is regenerative by design.[129] One component of this is cradle-to-cradle product design and manufacturing, also known as regenerative or sustainable design, most associated with the architect William McDonough and the chemist Michael Braungart.[130] The Cradle to Cradle Products Innovation Institute certifies consumer products, clothing, and building supplies according to their material health (safety and benefit for humans and the environment), material reutilization, renewable energy and carbon management, water stewardship, and social fairness.[131] Developing stronger social signals for these products, and consumption norms that link production to consumption more transparently, strikes us as a powerful way to better align consumption with social goals surrounding environmental justice.

A New Cool?

As much as we need a new paradigm of production, however, ultimately we likewise need a new paradigm of consumption. Much of the debate between consumerism and sustainability is still mired in traditional views of consumption.[132] Consumption remains "the problem," and recommendations to fix that problem rarely extend beyond the stock admonition that individuals should simply reduce their levels of consumption. The problem with such a recommendation is that on its own, reduced consumption creates a Status Dilemma. For this reason, people will resist it. A more effective way to change consumption behavior is to align socially beneficial consumption patterns with status norms, as we saw in the case of conspicuous conservation. Indeed, doing so has the capacity to create the sort of competitively altruistic dynamics we explored in chapter 4. In other words, rather than simply embrace a norm of reduced consumption, new consumption behaviors—such as those based on sustainable production technologies—could be aligned with status norms to more rapidly change consumption. What intrigues us most about new production paradigms such as Cradle to Cradle—and their role in "green growth," a "green economy," even a "Green New Deal"—is the potential to align them with consumption norms that grew out of DotCool. All of these approaches to sustainability, a post-carbon economy, and an era of post-scarcity made possible by emerging technologies are driven by the innovation and unconventionality norms we've highlighted in this chapter.[133] And all build on growing stronger links between consumption and morality, suggesting the potential of emerging norms to rapidly change our consumption behavior. To go beyond the stock explanations of our consumption, we'll need to recognize its biological underpinnings in the affiliative logic of social selection and the human need for status. We'll need to recognize the capacity of consumerism to solve the Status Dilemma and increase happiness. Finally, we'll need to recognize that a new paradigm of consumption and production not only can be aligned with our long-term social and environmental goals, but also has the potential to accelerate our realization of those goals by creating status incentives that tap into some of our most basic affiliative impulses. That would be cool.

NOTES

1. THE CONSUMPTION MYSTERY

1. A rich and contentious literature exists regarding the so-called birth of the consumer society and the emergence of the "consumer." These debates are tangential to the ones we're concerned with, which deal with the microfoundations of consumption. See, for example, McKendrick, Neil, John Brewer, and J. H. Plumb. 1982. *The Birth of a Consumer Society: The Commercialization of Eighteenth-Century England.* Bloomington: Indiana University Press. See also Clark, Gregory. 2010. "The Consumer Revolution: Turning Point in Human History, or Statistical Artifact?," working paper. www.econ.ucdavis.edu/faculty/gclark/papers/Consumer%20Revolution.pdf.
2. Sahadi, Jeanne. 2014. "How Much Do You Need to Be Happy?" CNN Money, June 5. http://money.cnn.com/2014/06/05/news/economy/how-much-income-to-be-happy/; CNN|ORC Poll. 2014. http://i.cdn.turner.com/money/2014/story-supplement/rel6c.pdf?iid=EL.
3. Hotz, Robert Lee. 2005. "Searching for the Why of Buy." *LA Times*, February 27.
4. Quartz, Steven R., and Terrence J. Sejnowski. 2002. *Liars, Lovers, and Heroes: What the New Brain Science Reveals About How We Become Who We Are.* New York: HarperCollins; Quartz and Sejnowski. 1997. "The neural basis of cognitive development: a constructivist manifesto." *Behavioral and Brain Sciences* 20:537–56; discussion 556–96.
5. By "the traditional economic model," we mean the standard textbook treatment of the consumer. As we see in later chapters, many economic studies now incorporate social identity.
6. See, for example, Teffer, Kate, Daniel P. Buxhoeveden, Cheryl D. Stimpson, Archibald J. Fobbs, Steven J. Schapiro, Wallace B. Baze, Mark J. McArthur, William D. Hopkins, Patrick R. Hof, Chet C. Sherwood, and Katerina Semendeferi. 2013. "Developmental changes in the spatial organization of neurons in the neocortex of humans and common chimpanzees." *Journal of*

Comparative Neurology 521:4249–59; Hill, Jason, Terrie Inder, Jeffrey Neil, Donna Dierker, John Harwell, and David Van Essen. 2010. "Similar patterns of cortical expansion during human development and evolution." *Proceedings of the National Academy of Sciences* 107:13135–40; Bianchi, Serena, Cheryl D. Stimpson, Tetyana Duka, Michael D. Larsen, William G. M. Janssen, Zachary Collins, Amy L. Bauernfeind, Steven J. Schapiro, Wallace B. Baze, Mark J. McArthur, William D. Hopkins, Derek E. Wildman, Leonard Lipovich, Christopher W. Kuzawa, Bob Jacobs, Patrick R. Hof, and Chet C. Sherwood. 2013. "Synaptogenesis and development of pyramidal neuron dendritic morphology in the chimpanzee neocortex resembles humans." *Proceedings of the National Academy of Sciences* 110 (Supplement 2):10395–401; Blakemore, Sarah-Jayne. 2008. "Development of the social brain during adolescence." *The Quarterly Journal of Experimental Psychology* 61:40–49; Mills, Kathryn L., François Lalonde, Liv S. Clasen, Jay N. Giedd, and Sarah-Jayne Blakemore. 2014. "Developmental changes in the structure of the social brain in late childhood and adolescence." *Social Cognitive and Affective Neuroscience* 9:123–31.

7. Godoy, Ricardo, et al. 2007. "Signaling by consumption in a native Amazonian society." *Evolution and Human Behavior* 28:124–34.

8. Baudrillard, Jean. 1970/1998. *The Consumer Society: Myths and Structures.* Trans. Chris Turner. London: Sage. Though we noted that Baudrillard was a postmodernist, this book, a relatively early work, is more modernist in tone.

9. De Botton, Alain. 2004. *Status Anxiety.* New York: Pantheon, p. 43.

10. Ibid, p. 44.

11. Cohen, Lizabeth. 2003. *A Consumers' Republic: The Politics of Mass Consumption in Postwar America.* New York: Alfred A. Knopf.

12. Herman, Edward, and Noam Chomsky. 1988. *Manufacturing Consent: The Political Economy of the Mass Media.* New York: Pantheon.

13. Although consumerism is most often linked to capitalism, the following studies illustrate the important interplay between consumerism and communism: Zatlin, Jonathan R. 2007. *The Currency of Socialism: Money and Political Culture in East Germany.* (Publications of the German Historical Institute.) Cambridge, UK: Cambridge University Press; Landsman, Mark. 2005. *Dictatorship and Demand: The Politics of Consumerism in East Germany.* (Harvard Historical Studies.) Cambridge, MA: Harvard University Press; Gerth, Karl. 2013. "Compromising with consumerism in socialist China: Transnational flows and internal tensions in 'socialist advertising.'" *Past & Present* 218 (Supplement 8):203–32; Mazurek, M., and M. Hilton. 2007. "Consumerism, solidarity and communism: Consumer protection and the consumer movement in Poland." *Journal of Contemporary History* 42:315–43.

14. See the World Bank, "China Overview," www.worldbank.org/en/country/china/overview.

15. Barber, Benjamin R. 1995. *Jihad vs. McWorld: How the Planet Is Both Falling Apart and Coming Together and What This Means for Democracy.* New York: Crown.

16. Perhaps the most famous exception is Adam Smith, who viewed consumption as the end and purpose of production in *The Wealth of Nations* (1776).

17. This is sometimes described as a "Malthusian trap" after Thomas Malthus's 1798 work, "An Essay on the Principle of Population." See, for example, Clark,

Gregory. 2007. *A Farewell to Alms: A Brief Economic History of the World*. Princeton, NJ: Princeton University Press.

18. Weber, Max. 1905. *The Protestant Ethic and the Spirit of Capitalism*.

19. Lunbeck, Elizabeth. 2014. *The Americanization of Narcissism*. Cambridge, MA: Harvard University Press.

20. Lasch, Christopher. 1978. *The Culture of Narcissism: American Life in an Age of Diminishing Expectations*. New York: W. W. Norton. Narcissistic personality disorder has a prevalence of about 1 percent. The classification itself is highly contested. See, for example, the following regarding the status of narcissistic personality disorder: Schulze, Lars, and Stefan Roepke. 2014. "Structural and functional brain imaging in borderline, antisocial, and narcissistic personality disorder." In Christoph Mulert and Martha E. Shenton, eds. *MRI in Psychiatry*. Berlin: Springer Berlin Heidelberg, pp. 313–40; Paris, Joel. 2014. "After DSM-5: Where does personality disorder research go from here?" *Harvard Review of Psychiatry* 22:216–21.

21. Easterlin, Richard. 1974. "Does Economic Growth Improve the Human Lot? Some Empirical Evidence." In Paul A. David and Melvin W. Reder, eds. *Nations and Households in Economic Growth: Essays in Honor of Moses Abramovitz*. New York: Academic Press, pp. 89–125; Easterlin, Richard A. 1995. "Will raising the incomes of all increase the happiness of all?" *Journal of Economic Behavior and Organization* 27:35–47; Easterlin, R., A.L.A. McVey, M. Switek, O. Sawangfa, and J. S. Zweig. 2010. "The happiness–income paradox revisited." *Proceedings of the National Academy of Sciences* 107: 22463–68.

22. Frank, Robert H. 2011. *The Darwin Economy: Liberty, Competition, and the Common Good*. Princeton, NJ: Princeton University Press; Frank, Robert H. 1999. *Luxury Fever: Why Money Fails to Satisfy in an Era of Excess*. New York: Free Press.

23. Stevenson, Betsey, and Justin Wolfers. 2013. "Subjective well-being and income: Is there any evidence of satiation?" *American Economic Review* 103:598–604; Sacks, D. W., B. Stevenson, and J. Wolfers. 2012. "The new stylized facts about income and subjective well-being." *Emotion* 12:1181–87.

24. While our emphasis is on absolute wealth, we are aware of at least one economic model that shows how status concerns and relative position may increase happiness via a positive feedback process that contributes to economic growth and increased utility. See Strulik, H. 2013. "How Status Concerns Can Make Us Rich and Happy." Center for European Governance and Economic Development Research Discussion Papers no. 170.

25. Diener, Ed, Louis Tay, and Shigehiro Oishi. 2013. "Rising income and the subjective well-being of nations." *Journal of Personality and Social Psychology* 104:267–76.

26. Klein, Naomi. 2014. "Climate change is the fight of our lives—yet we can hardly bear to look at it." *The Guardian*, April 23. www.theguardian.com /commentisfree/2014/apr/23/climate-change-fight-of-our-lives-naomi-klein.

27. CBS News/New York Times Poll. 2006.

28. See, for example, Morewedge, Carey K. 2013. "It was a most unusual time: How memory bias engenders nostalgic preferences." *Journal of Behavioral Decision Making* 26:319–26.

29. Gray, John. 2011. "Delusions of Peace." *Prospect*, September 21. www .prospectmagazine.co.uk/features/john-gray-steven-pinker-violence-review.

30. 2014 Gates Annual Letter, http://annualletter.gatesfoundation.org/.

31. Hans Rosling's *200 Countries, 200 Years, 4 Minutes*. BBC. 2010. www.youtube .com/watch?v=jbkSRLYSojo.

32. See, however, Brennan, Geoffrey, and Philip Pettit. 2004. *The Economy of Esteem: An Essay on Civil and Political Society*. Oxford, UK: Oxford University Press.

2. THE THREE CONSUMERS WITHIN

1. See, for example, Dayan, Peter, Yael Niv, Ben Seymour, and Nathaniel D. Daw. 2006. "The misbehavior of value and the discipline of the will." *Neural Networks* 19:1153–60; Rangel, Antonio, Colin Camerer, and P. Read Montague. 2008. "A framework for studying the neurobiology of value-based decision making." *Nature Reviews Neuroscience* 9:545–56.

2. Survival and Habit systems can guide social behavior as well, but we have in mind a particular sort of social behavior, directed by the Goal system, that involves computing the expected utility of options in terms of their personal social valuation.

3. Although this is a slightly fanciful way of putting it, in fact the link between defiant music and rebellious youth is well established. See, for example, Carpentier, Francesca D., Silvia Knobloch, and Dolf Zillmann. 2003. "Rock, rap, and rebellion: Comparisons of traits predicting selective exposure to defiant music." *Personality and Individual Differences* 35:1643–55.

4. Belfiore, Elizabeth S. 2006. "Dancing with the gods: The myth of the chariot in Plato's 'Phaedrus.'" *The American Journal of Philology* 127:185–217.

5. See, for example, MacDonald, K. 1986. "Civilization and Its Discontents Revisited: Freud as an evolutionary biologist." *Journal of Social and Biological Structures* 9:307–18; Rodgers, Joann Ellison. 2001. *Sex: A Natural History*. New York: W. H. Freeman; for a more philosophical critique, see Grünbaum, Adolf. 1984. *The Foundations of Psychoanalysis: A Philosophical Critique*. Berkeley: University of California Press.

6. Freud is obviously not the only developmental theorist to posit developmental stages. Perhaps the best-known such account was that of Jean Piaget, although the notion of discrete and well-ordered stages proved problematic.

7. Robson, Shannen L., and Bernard Wood. 2008. "Hominin life history: Reconstruction and evolution." *Journal of Anatomy* 212:394–425.

8. See, for example, Bianchi, Serena, Cheryl D. Stimpson, Tetyana Duka, Michael D. Larsen, William G. M. Janssen, Zachary Collins, Amy L. Bauernfeind, Steven J. Schapiro, Wallace B. Baze, Mark J. McArthur, William D. Hopkins, Derek E. Wildman, Leonard Lipovich, Christopher W. Kuzawa, Bob Jacobs, Patrick R. Hof, and Chet C. Sherwood. 2013. "Synaptogenesis and development of pyramidal neuron dendritic morphology in the chimpanzee neocortex resembles humans." *Proceedings of the National Academy of Sciences* 110 (Supplement 2):10395–401; Teffer, Kate, Daniel Buxhoeveden, Cheryl Stimpson, Archibald Fobbs, Steven Schapiro, Wallace Baze, Mark McArthur, William

Hopkins, Patrick Hof, Chet Sherwood, and Katerina Semendeferi. 2013. "Developmental changes in the spatial organization of neurons in the neocortex of humans and common chimpanzees." *Journal of Comparative Neurology* 521:4249–59.

9. Petanjek, Zdravko, Miloš Judaš, Goran Šimić, Mladen Roko Rašin, Harry B. M. Uylings, Pasko Rakic, and Ivica Kostović. 2011. "Extraordinary neoteny of synaptic spines in the human prefrontal cortex." *Proceedings of the National Academy of Sciences* 108:13281–86.

10. Haider, Aliya. 2006. "Roper v. Simmons: The role of the science brief." *Ohio State Journal of Criminal Law* 3:369–77.

11. Quartz, Steven R., and Terrence J. Sejnowski. 2002. *Liars, Lovers, and Heroes: What the New Brain Science Reveals About How We Become Who We Are.* New York: William Morrow.

12. Haidt, Jonathan, S. H. Koller, and M. G. Dias. 1993. "Affect, culture, and morality, or Is it wrong to eat your dog?" *Journal of Personality and Social Psychology* 65:613–28.

13. Despite the emphasis on flourishing in Aristotle, his conception of happiness meant the activity of intelligence in the form of contemplation. See Gurtler, Gary M. 2003. "The activity of happiness in Aristotle's Ethics." *Review of Metaphysics* 56:801–34.

14. Bentham, Jeremy. 1789. *An Introduction to the Principles of Morals and Legislation.*

15. The term "utility," however, had been used before Bentham, such as in the writings of David Hume and Adam Smith. See, for example, Raphael, D. D. 1972. "Hume and Adam Smith on Justice and Utility." *Proceedings of the Aristotelian Society*, New Series 73:87–103.

16. Edgeworth, Francis Ysidro. 1881. *Mathematical Psychics: An Essay on the Application of Mathematics to the Moral Sciences.* London: Kegan Paul.

17. For a historical perspective, see McFadden, Daniel L. 2013. "The New Science of Pleasure." NBER Working Paper Series no. 18687; Colander, David. 2007. "Retrospectives: Edgeworth's hedonimeter and the quest to measure utility." *Journal of Economic Perspectives* 21:215–26.

18. Even more simply, if you are presented with a choice between two goods, then the option you choose will have more utility for you than the option you'd pass over. This latter sense of utility is known as ordinal utility in contrast to cardinal utility. A cardinal utility measurement allows one to measure the amount of utility, roughly akin to the way we measure temperature in degrees. In contrast, an ordinal measurement only allows one to think quantitatively about utility in such terms as "greater than" or "less than." Ordinal utility finally won the day, and the idea that utility was a subjective experience of pleasure was eventually abandoned. If economics could do with the weaker ordinal conception of utility, then the measurement problem was much easier, all the while abstracting further and further away from the physiological foundations of utility.

19. Olds, James, and Peter Milner. 1954. "Positive reinforcement produced by electrical stimulation of septal area and other regions of rat brain." *Journal of Comparative and Physiological Psychology* 47:419–27.

20. Olds, James. 1956. "Pleasure Centers in the Brain." *Scientific American* 195:105–17.

21. Wise, Roy A. 2008. "Dopamine and reward: The anhedonia hypothesis 30 years on." *Neurotoxicity Research* 14:169–83.

22. Schultz, W., P. Apicella, and T. Ljungberg. 1993. "Responses of monkey dopamine neurons to reward and conditioned stimuli during successive steps of learning a delayed response task." *Journal of Neuroscience* 13:900–913.

23. Montague, P. Read, Peter Dayan, Christophe Person, and Terrence J. Sejnowski. 1995. "Bee foraging in uncertain environments using predictive Hebbian learning." *Nature* 377:725–28; Montague, P. R., P. Dayan, and T. J. Sejnowski. 1996. "A framework for mesencephalic dopamine systems based on predictive Hebbian learning." *Journal of Neuroscience* 16:1936–47; W. Schultz, P. Dayan, and P. R. Montague. 1997. "A neural substrate of prediction and reward." *Science* 275:1593–99.

24. Some neuroscientists, such as Paul Glimcher at NYU, were also beginning to use economic theory to help guide their experiments, including ones involving how animals generate reward-related behavior. See Glimcher, Paul W. 2003. *Decisions, Uncertainty, and the Brain: The Science of Neuroeconomics.* Cambridge, MA: MIT Press.

25. Prominent among these were behavioral economists, including Colin Camerer at Caltech, George Loewenstein at Carnegie Mellon, and Drazen Prelec at MIT, who cowrote a neuroeconomic manifesto of sorts in 2005. See Camerer, Colin, George Loewenstein, and Drazen Prelec. 2005. "Neuroeconomics: How neuroscience can inform economics." *Journal of Economic Literature* 43:9–64.

26. Preuschoff, Kerstin, Peter Bossaerts, and Steven R. Quartz. 2005. "Neural differentiation of expected reward and risk in human subcortical structures." *Neuron* 51:381–90; Preuschoff, Kerstin, Steven R. Quartz, and Peter Bossaerts. 2008. "Human insula activation reflects risk prediction errors as well as risk." *The Journal of Neuroscience* 28:2745–52.

27. Bruguier, Antoine, Kerstin Preuschoff, Steven Quartz, and Peter Bossaerts. 2008. "Investigating signal integration with canonical correlation analysis of fMRI brain activation data." *NeuroImage* 41:35–44.

28. This is closely related to predicted utility, but is emphasized by economists because it is the utility that is revealed by our decisions and so plays a special role in revealed preference theory.

29. Redelmeier, Donald A., Joel Katz, and Daniel Kahneman. 2003. "Memories of colonoscopy: A randomized trial." *Pain* 104:187–94.

30. Beierholm, Ulrik R., Cédric Anen, Steven Quartz, and Peter Bossaerts. 2011. "Separate encoding of model-based and model-free valuations in the human brain." *NeuroImage* 58:955–62.

31. While sometimes the description of the brain as a computer is taken to be a metaphor, the actual claim is much stronger: the brain is a computational system. See, for example, Quartz, Steven R. 2009. "Reason, emotion and decision-making: Risk and reward computation with feeling." *Trends in Cognitive Sciences* 13:209–15.

32. Tusche, Anita, Stefan Bode, and John-Dylan Haynes. 2010. "Neural responses to unattended products predict later consumer choices." *The Journal of Neuroscience* 30:8024–31.

33. This is a complex issue. For more information, see Quartz, S. R. 2003. "Innateness and the brain." *Biology and Philosophy* 18:13–40; Quartz, S. R. 1993. "Neural networks, nativism, and the plausibility of constructivism." *Cognition* 48:223–42.

34. Bushong, Benjamin, Lindsay M. King, Colin F. Camerer, and Antonio Rangel. "Pavlovian processes in consumer choice: The physical presence of a good increases willingness-to-pay." *American Economic Review* 100:1–18.

35. "Grocery cart choice architecture." 2010. *Nudge* (blog), August 13. http://nudges .org/2010/08/13/grocery-cart-choice-architecture/.

36. Wengrow, David. 2008. "Prehistories of commodity branding." *Current Anthropology* 49:7–34.

37. McClure, S. M., J. Li, D. Tomlin, K. S. Cypert, L. M. Montague, and P. R. Montague. 2004. "Neural correlates of behavioral preference for culturally familiar drinks." *Neuron* 44:379–87.

38. Koenigs, Michael, and Daniel Tranel. 2008. "Prefrontal cortex damage abolishes brand-cued changes in cola preference." *Social Cognitive and Affective Neuroscience* 3:1–6.

39. Bhasin, Kim. 2013. "Lululemon Admits Its PR Disasters Are Hurting Sales." *Huffington Post*, December 12. www.huffingtonpost.com/2013/12/12/lululemon -pr_n_4434580.html.

40. Dijksterhuis, A., et al. 2006. "On making the right choice: The deliberation-without-attention effect." *Science* 311:1005–07.

41. O'Doherty, John P., Tony W. Buchanan, Ben Seymour, and Raymond J. Dolan. 2006. "Predictive neural coding of reward preference involves dissociable responses in human ventral midbrain and ventral striatum." *Neuron* 49:157–66.

42. Olenski, Steve. 2013. "Is Brand Loyalty Dying a Slow and Painful Death?" *Forbes*, January 7. www.forbes.com/sites/marketshare/2013/01/07/is-brand -loyalty-dying-a-slow-and-painful-death/.

3. COOL ON THE BRAIN

1. "Coolquest." www.nbcnews.com/id/5377665/ns/msnbc-national_geographic _ultimate_explorer/t/coolquest/#.U_i8NfldV8E.

2. Lattman, Peter. 2007. "The Origins of Justice Stewart's 'I Know It When I See It.'" *The Wall Street Journal*, September 27. http://blogs.wsj.com/law/2007/09/27 /the-origins-of-justice-stewarts-i-know-it-when-i-see-it/.

3. Gladwell, Malcolm. 1997. "The Coolhunt." *The New Yorker*, March 17.

4. Luckerson, Victor. 2013. "Is Facebook Losing Its Cool? Some Teens Think So." *Time*, March 8. http://business.time.com/2013/03/08/is-facebook-losing-its -cool-some-teens-think-so/.

5. Tabuchi, Hiroko. 2013. "Sony's Bread and Butter? It's Not Electronics." *The New York Times*, May 27. www.nytimes.com/2013/05/28/business/global/sonys -bread-and-butter-its-not-electronics.html.

6. Jabr, Ferris. 2012. "Does Thinking Really Hard Burn More Calories?" *Scientific American*, July 18. www.scientificamerican.com/article/thinking-hard-calories/.

7. Komisaruk, Barry R., and Beverly Whipple. 2005. "Functional MRI of the brain during orgasm in women." *Annual Review of Sex Research* 16:62–86.

8. Semendeferi, Katerina, and Hanna Damasio. 2000. "The brain and its main anatomical subdivisions in living hominoids using magnetic resonance imaging." *Journal of Human Evolution* 38:317–332.

9. Semendeferi, K., et al. 2001. "Prefrontal cortex in humans and apes: A comparative study of area 10." *American Journal of Physical Anthropology* 114:224–41.

10. Rosenbaum, R. Shayna, Stefan Köhler, Daniel L. Schacter, Morris Moscovitch, Robyn Westmacott, Sandra E. Black, Fuqiang Gao, and Endel Tulving. 2005. "The case of K.C.: Contributions of a memory-impaired person to memory theory." *Neuropsychologia* 43:989–1021.

11. Reviewed in Northoff, Georg, and Felix Bermpohl. 2004. "Cortical Midline Structures and the Self." *Trends in Cognitive Sciences* 8:102–07; and Wagner, D. D., J. V. Haxby, and T. F. Heatherton. 2012. "The Representation of Self and Person Knowledge in the Medial Prefrontal Cortex." *WIREs Cognitive Science* 3:451–70.

12. This idea of the development of the self through social feedback is closely associated with Charles Cooley's "looking-glass self," a theory he developed in the early twentieth century. Indeed, one of the most intriguing insights about the MPFC is its protracted development, which aligns with the protracted development of the human self-concept.

13. Weise, Elizabeth. 2010. "93% of Women Wash Their Hands vs. 77% of men." *USA Today*, September 13. http://usatoday30.usatoday.com/yourlife/health/2010 -09-13-handwashing14_ST_N.htm?csp=34news.

14. See also the work of Mark Leary on the sociometer. Leary, M. R., and R. F. Baumeister. 2000. "The nature and function of self-esteem: Sociometer theory." In M. P. Zanna, ed., *Advances in Experimental Social Psychology*, Vol. 32, pp. 1–62. San Diego, CA: Academic Press.

15. Somerville, Leah H., et al. 2013. "The medial prefrontal cortex and the emergence of self-conscious emotion in adolescence." *Psychological Science* 24: 1554–62.

16. Eisenberger, Naomi I., et al. 2011. "The neural sociometer: Brain mechanisms underlying state self-esteem." *Journal of Cognitive Neuroscience* 23:3448–55.

17. Baumeister, Roy F., et al. 2001. "Bad is stronger than good." *Review of General Psychology* 5:323–70.

18. Izuma, Keise, Daisuke N. Saito, and Norihiro Sadato. 2010. "The roles of the medial prefrontal cortex and striatum in reputation processing." *Social Neuroscience* 5:133–47.

19. Davey, Christopher G., et al. 2010. "Being liked activates primary reward and midline self-related brain regions." *Human Brain Mapping* 31:660–68.

20. Izuma et al., 2010, "Roles of the medial prefrontal cortex and striatum."

21. Lin, Alice, Ralph Adolphs, and Antonio Rangel. 2012. "Social and monetary reward learning engage overlapping neural substrates." *Social Cognitive and Affective Neuroscience* 7:274–81.

22. Leary, Mark R. 2007. "Motivational and emotional aspects of the self." *Annual Review of Psychology* 58:317–44.

23. Bushman, Brad J., Scott J. Moeller, and Jennifer Crocker. 2011. "Sweets, sex, or self-esteem? Comparing the value of self-esteem boosts with other pleasant rewards." *Journal of Personality* 79:993–1012.

24. Baumeister, 2001, "Bad is stronger than good."
25. Berthoz, S., et al. 2002. "An fMRI study of intentional and unintentional (embarrassing) violations of social norms." *Brain* 125:1696–708.
26. Preuschoff, Quartz, and Bossaerts, 2008. "Human insula activation."
27. Henrich, Joseph, Robert Boyd, Samuel Bowles, Colin Camerer, Ernst Fehr, Herbert Gintis, Richard McElreath, Michael Alvard, Abigail Barr, Jean Ensminger, Natalie Smith Henrich, Kim Hill, Francisco Gil-White, Michael Gurven, Frank W. Marlowe, John Q. Patton, and David Tracer. 2005. "Economic man" in cross-cultural perspective: Behavioral experiments in 15 small-scale societies." *Behavioral and Brain Sciences* 28:795–815.
28. Montague, P. Read, and Terry Lohrenz. 2007. "To detect and correct: Norm violations and their enforcement." *Neuron* 56:14–8.
29. The BIS/BAS systems are most associated with Jeffrey Gray. See Gray, J. A. 1970. "The psychophysiological basis of introversion-extraversion." *Behaviour Research & Therapy*, 8:249–66; and Gray, J. A. 1990. "Brain systems that mediate both emotion and cognition." *Cognition and Emotion* 4:269–88.
30. Simon, Joe J., et al. 2010. "Neural reward processing is modulated by approach- and avoidance-related personality traits." *NeuroImage* 49:1868–74.
31. Canli, Turhan. 2004. "Functional brain mapping of extraversion and neuroticism: Learning from individual differences in emotion processing." *Journal of Personality* 72:1105–32.
32. Claes, Laurence, et al. 2010. "Emotional reactivity and self-regulation in relation to compulsive buying." *Personality and Individual Differences* 49:526–30.
33. Blair, Karina S., et al. 2010. "Social norm processing in adult social phobia: Atypically increased ventromedial frontal cortex responsiveness to unintentional (embarrassing) transgressions." *American Journal of Psychiatry* 167: 1526–32.

4. CONSUMER EVOLUTION

1. The sweater, charitable donation, and branded interview studies are reported in Nelissen, Rob M. A., and Marijn H. C. Meijers. 2011. "Social benefits of luxury brands as costly signals of wealth and status." *Evolution and Human Behavior* 32:343–55.
2. Bourdieu, Pierre. 1984. *Distinction: A Social Critique of the Judgement of Taste.* Cambridge, MA: Harvard University Press.
3. Sivanathan, Niro, and Nathan C. Pettit. 2010. "Protecting the self through consumption: Status goods as affirmational commodities." *Journal of Experimental Social Psychology* 46:564–70.
4. Rindfleisch, Aric, James E. Burroughs, and Nancy Wong. 2009. "The safety of objects: Materialism, existential insecurity, and brand connection." *Journal of Consumer Research* 36:1–16.
5. Solomon, Sheldon, Jeffrey L. Greenberg, and Thomas A. Pyszczynski. 2004. "Lethal consumption: Death-denying materialism." In T. Kasser and A. D. Kanner, eds. *Psychology and Consumer Culture: The Struggle for a Good Life in a Materialistic World.* Washington, DC: American Psychological Association, 127–46.

6. The $5 amount is based on an endowment of $20. For a review, see Engel, Christoph. 2011. "Dictator games: A meta study." *Experimental Economics* 14:583–610.

7. Gneezy, Uri, Ernan Haruvy, and Hadas Yafe. 2004. "The inefficiency of splitting the bill." *The Economic Journal* 114:265–80.

8. Daly, M., M. Wilson, C. A. Salmon, M. Hiraiwa-Hasegawa, and T. Hasegawa. 2001. "Siblicide and seniority." *Homicide Studies* 5:30–45.

9. Zerjal, Tatiana, et al. 2003. "The genetic legacy of the Mongols." *American Journal of Human Genetics* 72:717–21.

10. We often think of kin selection as the idea that we're altruistic to people in proportion to the genes we share, but it concerns a specific gene or genes relating to altruism. See Frank, S. A. 2013. "Natural selection. VII. History and interpretation of kin selection theory." *Journal of Evolutionary Biology* 26:1151–84.

11. Seyfarth, Robert M., and Dorothy L. Cheney. 2012. "The evolutionary origins of friendship." *Annual Review of Psychology* 63:153–77.

12. Silk, Joan B. 2003. "Cooperation without counting: The puzzle of friendship." In P. Hammerstein, ed. *Genetic and Cultural Evolution of Cooperation*. Dahlem Workshop Reports. Cambridge, MA: MIT Press, pp. 37–54.

13. Nesse, Randolph. 2007. "Runaway social selection for displays of partner value and altruism." *Biological Theory* 2:143–55.

14. See, for example, Miller, Geoffrey F. 2007. "Sexual selection for moral virtues." *The Quarterly Review of Biology* 82:97–125.

15. Kuhlmeier, Valerie, Karen Wynn, and Paul Bloom. 2003. "Attribution of dispositional states by 12-month-olds." *Psychological Science* 14:402–408.

16. Barclay, Pat, and Robb Willer. 2007. "Partner choice creates competitive altruism in humans." *Proceedings of the Royal Society B: Biological Sciences* 274: 749–53.

17. Signaling theory derives from two main sources, an economic perspective and a biological one. A major difference between these views stems from the fact that the economic view typically regards signals to be conscious elements of strategic interactions, as when an employer views an undergraduate degree in liberal arts from an Ivy League school as a signal that a prospective employee would be a hard worker (as opposed to a candidate with specific job-related skills, such as an MBA would signal). In this case, the signaling quality of the degree is consciously recognized by both employer and prospective employee, who, the economic view assumes, pursued the degree in part because of her conscious recognition of its signaling function in the job market. See Spence, A. M. (1976). "Competition in salaries, credentials, and signaling prerequisites for jobs." *The Quarterly Journal of Economics* 90:51–74.

18. Cottrell, Catherine A., Steven L. Neuberg, and Norman P. Li. 2007. "What do people desire in others? A sociofunctional perspective on the importance of different valued characteristics." *Journal of Personality and Social Psychology* 92:208–31.

19. Leary, Mark R. 2007. "Motivational and emotional aspects of the self." *Annual Review of Psychology* 58:317–44.

20. Wilson, Rick K., and Catherine C. Eckel. 2006. "Judging a book by its cover: Beauty and expectations in the trust game." *Political Research Quarterly* 59:189–202.

21. DeBruine, Lisa M. 2002. "Facial resemblance enhances trust." *Proceedings of the Royal Society B: Biological Sciences* 269:1307–12.

22. Stirrat, M., and D. I. Perrett. 2010. "Valid facial cues to cooperation and trust: Male facial width and trustworthiness." *Psychological Science* 21:349–54.

23. Carré, Justin M., and Cheryl M. McCormick. 2008. "In your face: Facial metrics predict aggressive behaviour in the laboratory and in varsity and professional hockey players." *Proceedings of the Royal Society B: Biological Sciences* 275:2651–56.

24. Carré, Justin M., Cheryl M. McCormick, and Catherine J. Mondloch. 2009. "Facial structure is a reliable cue of aggressive behavior." *Psychological Science* 20:1194–98.

25. Reviewed in Mende-Siedlecki, Peter, Christopher P. Said, and Alexander Todorov. 2013. "The social evaluation of faces: A meta-analysis of functional neuroimaging studies." *Social Cognitive and Affective Neuroscience* 8:285–99.

26. Engell, Andrew D., James V. Haxby, and Alexander Todorov. 2007. "Implicit trustworthiness decisions: Automatic coding of face properties in the human amygdala." *Journal of Cognitive Neuroscience* 19:1508–19.

27. Duarte, J., S. Siegel, and L. Young. 2012. "Trust and credit: The role of appearance in peer-to-peer lending." *Review of Financial Studies* 25:2455–84.

28. "Cone-ing Is the New Planking!!!!!" www.youtube.com/watch?v=WygNjMSllLQ.

29. "Cone-ing Same McDonalds 3 Times in a Row." www.youtube.com/watch?v=_JLnjq4uoJ0.

30. Bicchieri, C. 2005. *The Grammar of Society: The Nature and Dynamics of Social Norms.* Cambridge, UK: Cambridge University Press.

31. Posner, Eric A. 2000. *Law and Social Norms.* Cambridge, MA: Harvard University Press.

32. Cottrell, Catherine A., Steven L. Neuberg, and Norman P. Li. 2007. "What do people desire in others? A sociofunctional perspective on the importance of different valued characteristics." *Journal of Personality and Social Psychology* 92:208–31.

33. D'Errico, Francesco, et al. 2009. "Additional evidence on the use of personal ornaments in the Middle Paleolithic of North Africa." *Proceedings of the National Academy of Sciences* 106:16051–56.

34. Charles, Kerwin Kofi, Erik Hurst, and Nikolai Roussanov. 2009. "Conspicuous Consumption and Race." *The Quarterly Journal of Economics* 124:425–67.

35. Anderson, Cameron, Robb Willer, Gavin J. Kilduff, and Courtney E. Brown. 2012. "The origins of deference: When do people prefer lower status?" *Journal of Personality and Social Psychology* 102:1077–88.

36. Simpson, B. 2006. "Social identity and cooperation in social dilemmas." *Rationality and Society* 18:443–70; Van Bavel, Jay J., Dominic J. Packer, and William A. Cunningham. 2008. "The neural substrates of in-group bias: A functional magnetic resonance imaging investigation." *Psychological Science* 19:1131–39.

5. STATUS SEEKING AND THE REBEL INSTINCT

1. Schouten, John W. and James H. McAlexander. 1995. "Subcultures of consumption: An ethnography of the new bikers." *Journal of Consumer Research* 22: 43–61.

2. Schouten, J. W., D. M. Martin, and J. H. McAlexander. 2007. "The Evolution of a Subculture of Consumption." In Bernard Cova, Robert V. Kozinets, and Avi Shankar, eds. *Consumer Tribes*. New York: Routledge, pp. 67–75.

3. See, for example, Miller, Daniel. 1987. *Material Culture and Mass Consumption*. New York: Basil Blackwell. For a review, see Sassatelli, Roberta. 2007. *Consumer Culture: History, Theory, and Politics*. Los Angeles: Sage.

4. Inglehart, Ronald. 1977. *The Silent Revolution: Changing Values and Political Styles Among Western Publics*. Princeton, NJ: Princeton University Press; Inglehart, Ronald F. 2008. "Changing values among Western publics from 1970 to 2006." *West European Politics* 31:130–46; Welzel, Christian, and Ronald Inglehart. 2010. "Agency, values, and well-being: A human development model." *Social Indicators Research* 97:43–63.

5. Inglehart refers to these new values as "postmaterialist" ones—we've referred to them as "post-scarcity" values (another term Inglehart uses) to avoid the confusing connotation postmaterialism has as potentially referring to values that reject material possessions. In contrast, post-scarcity assumes an abundance of material possessions, which play an increasing role in the satisfaction of postmaterialist goals.

6. While there is considerable controversy regarding what Thatcher meant by her remark, the explanatory shift itself was from macro-level explanations (i.e., macroeconomics) to accounts emphasizing individual actors and agency (i.e., microeconomics).

7. Creanza, N., L. Fogarty, and M. W. Feldman. 2012. "Models of cultural niche construction with selection and assortative mating." *PLoS ONE* 7:e42744.

8. Jost, John T., Vagelis Chaikalis-Petritsis, Dominic Abrams, Jim Sidanius, Jojanneke van der Toorn, and Christopher Bratt. 2012. "Why men (and women) do and don't rebel: Effects of system justification on willingness to protest." *Personality and Social Psychology Bulletin* 38:197–208.

9. Kandler, Christian, Wiebke Bleidorn, and Rainer Riemann. 2012. "Left or right? Sources of political orientation: The roles of genetic factors, cultural transmission, assortative mating, and personality." *Journal of Personality and Social Psychology* 102:633–45; Kanai, Ryota, Tom Feilden, Colin Firth, and Geraint Rees. 2011. "Political orientations are correlated with brain structure in young adults." *Current Biology* 21:677–80.

10. Smith, Eric Alden, et al. 2010. "Production systems, inheritance, and inequality in premodern societies." *Current Anthropology* 51:85–94.

11. We're following the convention of related literature and calling these contests "zero-sum." Almost all purported examples of a zero-sum game in this literature aren't zero-sum (e.g., Prisoner's Dilemma).

12. For a review and defense, see Schor, J. B. 2007. "In defense of consumer critique: Revisiting the consumption debates of the twentieth century." *The Annals of the American Academy of Political and Social Science* 611:16–30.

13. Bourdieu, 1984, *Distinction*.

14. De Waal, Frans B. M. 1986. "The brutal elimination of a rival among captive male chimpanzees." *Ethology and Sociobiology* 7:237–51.

15. Smith, Heather J., et al. 2012. "Relative deprivation: A theoretical and meta-analytic review." *Personality and Social Psychology Review* 16:203–32.

16. Inglehart, Ronald, et al. 2008. "Development, freedom, and rising happiness." *Perspectives on Psychological Science* 3:264–85.

17. For a review of this idea, see Powell, Walter W., and Kaisa Snellman. 2004. "The knowledge economy." *Annual Review of Sociology* 30:199–220.

18. Bielby, William T. 2004. "Rock in a hard place: Grassroots cultural production in the post-Elvis era." *American Sociological Review* 69:1–13.

19. Bielby, "Rock in a hard place," 8.

20. United States Department of Labor, Bureau of Labor Statistics. http://data.bls .gov/timeseries/LNU04000000?years_option=all_years&periods_option= specific_periods&periods=Annual+Data.

21. Gair, Christopher. 2007. *The American Counterculture*. Edinburgh: Edinburgh University Press.

22. Rentfrow, Peter J., and Samuel D. Gosling. 2003. "The do re mi's of everyday life: The structure and personality correlates of music preferences." *Journal of Personality and Social Psychology* 84:1236–56.

23. Rentfrow, P. J., J. A. McDonald, and J. A. Oldmeadow. 2009. "You are what you listen to: Young people's stereotypes about music fans." *Group Processes & Intergroup Relations* 12:329–44.

24. Mulligan, Mark. 2014. "The Death of the Long Tail: The Superstar Music Economy." MIDiA Insights Report. Our estimate is based on listening to each song from start to end, assuming an average song length of four minutes. www .statcrunch.com/5.0/viewreport.php?reportid=28647&groupid=948.

25. This is, we believe, a conservative estimate, though the definition of a "genre" is open to debate.

26. Winograd, Morley, and Michael D. Hais. 2011. *Millennial Momentum: How a New Generation Is Remaking America*. New Brunswick, NJ: Rutgers University Press.

27. Milner states that "much of popular culture and scholarly research have drawn on a hierarchical imagery that obscures the more complex status relations in many (and perhaps most) high schools in a postmodern society." Milner, Murray, Jr. 2004. *Freaks, Geeks, and Cool Kids: American Teenagers, Schools, and the Culture of Consumption*. New York: Routledge, p. 127.

28. Milner, Murray. 2010. "Status Distinctions and Boundaries." In John R. Hall, Laura Grindstaff, and Ming Cheng Lo, eds. *Handbook of Cultural Sociology*. New York: Routledge, pp. 295–304.

29. Chen, Stephanie. 2010. "The 'Glee' effect: Singing is cool again." *CNN Living*, November 15. www.cnn.com/2010/LIVING/11/15/glee.effect.show.choir .comeback/.

30. Martin, Diane M., John W. Schouten, and James H. McAlexander. 2006. "Claiming the throttle: Multiple femininities in a hyper-masculine subculture." *Consumption, Markets and Culture* 9:171–205.

31. Boehm, Christopher. 2012. "Ancestral hierarchy and conflict." *Science* 336: 844–47.

32. See, for example, de Botton, Alain. 2004. *Status Anxiety*. New York: Pantheon.

33. Vanhaeren, Marian, and Francesco d'Errico. 2005. "Grave goods from the Saint-Germain-La-Rivière burial: Evidence for social inequality in the Upper Palaeolithic." *Journal of Anthropological Archaeology* 24:117–34.

34. Sassaman, Kenneth E. 2004. "Complex hunter-gatherers in evolution and history: A North American perspective." *Journal of Archaeological Research* 12: 227–80.

35. This is complicated by the finding that squirrel monkeys in captivity will stratify under certain conditions. See Bashaw, Meredith J., Chelsea McIntyre, and Nicole D. Salenetri. 2011. "Social organization of a stable natal group of captive Guyanese squirrel monkeys (*Saimiri sciureus sciureus*)." *Primates* 52:361–71.

36. Boehm, C. 1999. *Hierarchy in the Forest: The Evolution of Egalitarian Behavior.* Cambridge, MA: Harvard University Press; Boehm, C. 2012. *Moral Origins: The Evolution of Virtue, Altruism, and Shame.* New York: Basic Books.

37. Boehm, C. 2000. "Conflict and the evolution of social control." *Journal of Consciousness Studies* 7:79–101. Special Issue on Evolutionary Origins of Morality; Leonard Katz, guest editor.

38. Boehm, 2012. "Ancestral hierarchy and conflict."

39. Camerer, Colin F. 2003. *Behavioral Game Theory: Experiments in Strategic Interaction.* Princeton, NJ: Princeton University Press.

40. Sanfey, Alan G., James K. Rilling, Jessica A. Aronson, Leigh E. Nystrom, and Jonathan D. Cohen. 2003. "The neural basis of economic decision-making in the ultimatum game." *Science* 300:1755–58.

41. Brosnan, S. F., C. Talbot, M. Ahlgren, S. P. Lambeth, and S. J. Schapiro. 2010. "Mechanisms underlying responses to inequitable outcomes in chimpanzees, *Pan troglodytes*." *Animal Behaviour* 79:1229–37.

42. Ho, Arnold K., et al. 2012. "Social dominance orientation: Revisiting the structure and function of a variable predicting social and political attitudes." *Personality and Social Psychology Bulletin* 38:583–606.

43. Sapolsky, Robert M. 2004. "Social status and health in humans and other animals." *Annual Review of Anthropology* 33:393–418.

44. Marmot, Michael. 2004. *The Status Syndrome: How Social Standing Affects Our Health and Longevity.* New York: Times Books.

45. Chiao, Joan Y., Reginald B. Adams Jr., and Peter U. Tse. 2008. "Knowing who's boss: fMRI and ERP investigations of social dominance perception." *Group Process and Intergroup Relations* 11:201–14.

46. Kishida, Kenneth T., Dongni Yang, Karen Hunter Quartz, Steven R. Quartz, and P. Read Montague. 2012. "Implicit signals in small group settings and their impact on the expression of cognitive capacity and associated brain responses." *Philosophical Transactions of the Royal Society of London B: Biological Sciences* 367:704–16.

47. Yee, Vivian. 2013. "Grouping Students by Ability Regains Favor in Classroom." *The New York Times,* June 9, www.nytimes.com/2013/06/10/education /grouping-students-by-ability-regains-favor-with-educators.html?pagewanted =all&_r=0.

48. Godoy et al., 2007, "Signaling by consumption."

49. See chapter 1, note 1.

50. Smith, Eric Alden, Monique Borgerhoff Mulder, Samuel Bowles, Michael Gurven, Tom Hertz, and Mary K. Shenk. 2010. "Production systems, inheritance, and inequality in premodern societies." *Current Anthropology* 51:85–94; Bowles, Samuel, Eric Alden Smith, and Monique Borgerhoff Mulder. 2010.

"The emergence and persistence of inequality in premodern societies." *Current Anthropology* 51:7–17.

51. Smith, E. A., et al., 2010, "Production systems, inheritance, and inequality."
52. Boehm, 2012, "Ancestral hierarchy and conflict."
53. Ribeiro, A. 1986. *Dress and Morality*. London: Holmes & Meier.
54. Ribeiro, 1986, *Dress and Morality*.
55. Frank, R. H., 2011, *Darwin Economy*; Frank, 1999, *Luxury Fever*.
56. Kaburu, Stefano S. K., Sana Inoue, and Nicholas E. Newton-Fisher. 2013. "Death of the alpha: Within-community lethal violence among chimpanzees of the Mahale Mountains National Park." *American Journal of Primatology* 75:789–97.
57. Mitani, John C. 2009. "Cooperation and competition in chimpanzees: Current understanding and future challenges." *Evolutionary Anthropology* 18:215–27.
58. Mitani, 2009, "Cooperation and competition."
59. Hey, Jody. 2010. "The divergence of chimpanzee species and subspecies as revealed in multipopulation isolation-with-migration analyses." *Molecular Biology and Evolution* 27:921–33.
60. Hare, Brian, Victoria Wobber, and Richard Wrangham. 2012. "The self-domestication hypothesis: Evolution of bonobo psychology is due to selection against aggression." *Animal Behaviour* 83:573–85.
61. Watts, David P., et al. 2012. "Diet of chimpanzees (*Pan troglodytes schweinfurthii*) at Ngogo, Kibale National Park, Uganda, 1. Diet composition and diversity." *American Journal of Primatology* 74:114–29.
62. Robert G. Franciscus, et al. 2013. "Anatomically Modern Humans as a 'Self-domesticated' Species: Insights from Ancestral Wolves and Descendant Dogs." The 82nd Annual Meeting of the American Association of Physical Anthropologists.
63. The issue of warfare among early modern humans is particularly contentious. See Bowles, Samuel. 2009. "Did warfare among ancestral hunter-gatherers affect the evolution of human social behaviors?" *Science* 324:1293–98; Fry, Douglas P., and Patrik Söderberg. 2013. "Lethal aggression in mobile forager bands and implications for the origins of war." *Science* 341:270–73.
64. Henrich, Joseph, Robert Boyd, and Peter J. Richerson. 2012. "The puzzle of monogamous marriage." *Philosophical Transactions of the Royal Society of London B: Biological Sciences* 367:657–69.
65. Gray, Peter B. 2011. "The descent of a man's testosterone." *Proceedings of the National Academy of Sciences of the United States of America* 108:16141–42.

6. DARWIN GOES SHOPPING

1. Hector, Andy, and Rowan Hooper. 2002. "Ecology. Darwin and the first ecological experiment." *Science* 295:639–40.
2. Losos, Jonathan B. 2010. "Adaptive Radiation, Ecological Opportunity, and Evolutionary Determinism." American Society of Naturalists E. O. Wilson Award Address. *The American Naturalist* 175:623–39.
3. See, for example, Rainey, Paul B., and Michael Travisano. 1998. "Adaptive radiation in a heterogeneous environment." *Nature* 394:69–72; and Brockhurst,

Michael, Nick Colegrave, David J. Hodgson, and Angus Buckling. 2007. "Niche occupation limits adaptive radiation in experimental microcosms." *PloS One* 2:e193.

4. Sulloway, Frank J. 1996. *Born to Rebel: Birth Order, Family Dynamics, and Creative Lives*. New York: Pantheon.

5. Plomin, R., and D. Daniels. 1987. "Why are children in the same family so different from one another?" *Behavioral and Brain Sciences* 10:1–60.

6. "Eli Manning Talking Some Hoops and Some Pigskin." 2014. *Fox Sports Radio*, June 11. www.foxsportsradio.com/onair/jay-mohr-sports-50067/eli-manning-talking-some-hoops-and-12451118/.

7. Sulloway, 1996. *Born to Rebel*.

8. Hsu, Ming, Cédric Anen, and Steven R. Quartz. 2008. "The right and the good: Distributive justice and neural encoding of equity and efficiency." *Science* 320:1092–95.

9. Nozick, Robert. 1974. *Anarchy, State, and Utopia*. New York: Basic Books. p. 245.

10. Nozick, 1974, *Anarchy, State, and Utopia*, p. 246.

11. Epstein, David. 2013. *The Sports Gene: Inside the Science of Extraordinary Athletic Performance*. New York: Penguin.

12. Acemoglu, Daron, and James. A. Robinson. 2012. *Why Nations Fail: The Origins of Power, Prosperity, and Poverty*. New York: Crown Business.

13. Inglehart, Ronald, and Wayne E. Baker. 2000. "Modernization, cultural change, and the persistence of traditional values." *American Sociological Review* 65:19–51.

14. Jost et al., 2012, "Why men (and women) do and don't rebel."

15. Lee, I-Ching, F. Pratto, and B. T. Johnson. 2011. "Intergroup consensus/disagreement in support of group-based hierarchy: An examination of sociostructural and psycho-cultural factors." *Psychological Bulletin* 137:1029–64.

16. Barker, J. L., P. Barclay, and H. K. Reeve. 2012. "Within-group competition reduces cooperation and payoffs in human groups." *Behavioral Ecology* 23:735–41.

17. Henrich, Joseph, Robert Boyd, and Peter J. Richerson. 2012. "The puzzle of monogamous marriage." *Philosophical Transactions of the Royal Society of London B: Biological Sciences* 367:657–69.

18. Sherif, M., O. J. Harvey, B. J. White, W. R. Hood, and C. W. Sherif. 1954/1961. *The Robbers Cave Experiment: Intergroup Conflict and Cooperation.* Norman, OK: The University Book Exchange. pp. 15–18.

19. Van Vugt, Mark, David De Cremer, and Dirk P. Janssen. 2007. "Gender differences in cooperation and competition: The male-warrior hypothesis." *Psychological Science* 18:19–23.

20. Petersen, Roger. 2012. "Identity, Rationality, and Emotion in the Processes of State Disintegration and Reconstruction." In Kanchan Chandra, ed. *Constructivist Theories of Ethnic Politics*. Oxford, UK: Oxford University Press, pp. 387–421.

21. There is an interesting link between this finding and Robert Frank's claim that people compare themselves to others near them on the status hierarchy. For example, Frank often paraphrases Bertrand Russell, who notes that a beggar

doesn't envy the millionaire. He envies the beggar who is doing a bit better than he is. See Frank, R. H., 1999, *Luxury Fever.*

22. Examples during this period include *Brown v. Board of Education* in 1954; the Civil Rights Acts of 1957, 1964, and 1968; the 24th Amendment in 1964; the 1964 Voting Rights Act; and the Fair Housing Act and Equal Employment Opportunity Act in 1971.

23. Welzel and Inglehart, 2010, "Agency, values, and well-being."

24. Inglehart et al., 2008, "Development, freedom, and rising happiness."

25. Gair, 2007, *American Counterculture.*

26. For a succinct summary of these issues, see Oakes, Michael. 2012. "Measuring Socioeconomic Status." NIH Office of Behavioral and Social Science Research. E-source: Behavioral and Social Sciences Research (online textbook). www .esourceresearch.org/tabid/767/default.aspx.www.

27. Norton, Michael I., and Dan Ariely. 2011. "Building a better America—one wealth quintile at a time." *Perspectives on Psychological Science* 6:9–12.

28. Collins, Randall. 2000. "Situational stratification: A micro-macro theory of inequality." *Sociological Theory* 18:17–43.

29. Goffman, Erving. 1967. *Interaction Ritual: Essays on Face-to-Face Behavior.* New York: Doubleday.

30. Collins, 2000, "Situational stratification."

31. Wolff, L.S., D. Acevedo-Garcia, S. V. Subramanian, D. Weber, and I. Kawachi. 2010. "Subjective social status, a new measure in health disparities research: Do race/ethnicity and choice of referent group matter?" *Journal of Health Psychology* 15:560–74; Nobles, Jenna, Miranda Ritterman Weintraub, and Nancy E. Adler. 2013. "Subjective socioeconomic status and health: Relationships reconsidered." *Social Science & Medicine* 82:58–66.

32. Wolff et al., 2010, "Subjective social status."

33. Harris, Angel L. 2010. "The economic and educational state of black Americans in the 21st century: Should we be optimistic or concerned?" *The Review of Black Political Economy* 37:241–52; Wolff et al., 2010, "Subjective social status."

34. Frank, R. H., 1999, *Luxury Fever.*

35. See, for example, Bagwell, Laurie Simon, and B. Douglas Bernheim. 1996. "Veblen effects in a theory of conspicuous consumption." *American Economic Review,* 86:349–73; Hopkins, E., and T. Kornienko. 2004. "Running to keep in the same place: Consumer choice as a game of status." *American Economic Review* 94:1085–1107; Becker, G. S., K. M. Murphy, and I. Werning. 2005. "The equilibrium distribution of income and the market for status." *Journal of Political Economy* 113:282–310; Heffetz, Ori. 2012. "Who sees what? Demographics and the visibility of consumer expenditures." *Journal of Economic Psychology* 33:801–18.

36. Chernow, Ron. 1998. *Titan: The Life of John D. Rockefeller, Sr.* New York: Random House.

37. Asabere, Paul, and Forrest Huffman. 2013. "The impact of relative size on home values." *The Appraisal Journal,* p. 24.

38. Anderson, Cameron, and Gavin J. Kilduff. 2009. "The pursuit of status in social groups." *Current Directions in Psychological Science* 18:295–98.

39. Hall, Jonathan, and Hongyi Li. 2012. "Why Isn't Conspicuous Consumption More Conspicuous?" no. 1899, www.hongyi.li/papers/Conspicuous.pdf; Lavie, Moshik, "Show Me the Money: Status, Cultural Capital, and Conspicuous Consumption." (October 1, 2008). Available at http://ssrn.com/abstract=1328292 or http://dx.doi.org/10.2139/ssrn.1328292.

40. Huang, Shiying. 2013. "Bordeaux's Firsts Too Expensive for China." *The Drinks Business.* www.thedrinksbusiness.com/2013/07/bordeauxs-firsts-too-expensive -for-china/.

41. Smith, E. A., et al., 2010, "Production systems, inheritance, and inequality."

42. One of Max Weber's goals was to separate economic class from status, a distinction that is lost in many contemporary economic discussions where class is thought to unproblematically map isomorphically onto status. The fact that it does not is one reason why a society can be highly stratified economically but have little status stratification. See Chan, T. W., and J. H. Goldthorpe. 2007. "Class and status: The conceptual distinction and its empirical relevance." *American Sociological Review* 72:512–32.

43. Anderson, C., M. W. Kraus, A. D. Galinsky, and D. Keltner. 2012. "The local-ladder effect: Social status and subjective well-being." *Psychological Science* 23:764–71.

44. Wayland, Michael. 2012. "How crossovers, soccer moms 'killed' minivans." *MLive*, March 27–29. www.mlive.com/auto/index.ssf/2012/03/how_crossovers _soccer_moms_kil.html.

45. Bunkley, Nick. 2011. "Mocked as Uncool, the Minivan Rises Again." *The New York Times,* January 3. www.nytimes.com/2011/01/04/business/04minivan .html?pagewanted=all&_r=0.

46. Tversky, A., and E. Shafir. 1992. "Choice under conflict: The dynamics of deferred decision." *Psychological Science* 3:358–61.

47. Tversky and Shafir, 1992, "Choice under conflict."

48. Sweeny, Kate, and Kathleen D. Vohs. 2012. "On near misses and completed tasks: The nature of relief." *Psychological Science* 23:464–68.

49. Kohls, Gregor, et al. 2013. "The nucleus accumbens is involved in both the pursuit of social reward and the avoidance of social punishment." *Neuropsychologia* 51:2062–69.

50. White, Katherine, and Darren W. Dahl. 2007. "Are all out-groups created equal? Consumer identity and dissociative influence." *Journal of Consumer Research* 34:525–36.

51. See, for example, Goodman, Douglas J., and Mirelle Cohen. 2003. *Consumer Culture: A Reference Handbook.* Santa Barbara, CA: ABC-CLIO.

52. Escalas, J. E., and J. R. Bettman. 2005. "Self-construal, reference groups, and brand meaning." *Journal of Consumer Research* 32:378–89.

53. Bunkley, 2011, "Mocked as Uncool, the Minivan . . . "

54. Baskin, Jonathan S. 2013. "Harley-Davidson Will Be a Case Study in Social Branding." *Forbes*, July 12. www.forbes.com/sites/jonathansalembaskin/2013 /07/12/harley-davidson-will-be-a-case-history-in-social-branding/.

55. Cellini, Adelia. 2004. "The Story Behind Apple's '1984' TV Commercial: Big Brother at 20." *MacWorld* 21:18.

56. For a view related to identity goods, see Sunstein, Cass, and Edna Ullmann Margalit. 2001. "Solidarity Goods." *The Journal of Political Philosophy* 9:129–49.

57. Berger, Jonah A., and Chip Heath. 2007. "Where consumers diverge from others: Identity signaling and product domains." *Journal of Consumer Research* 34:121–34. Berger, Jonah, and Chip Heath. 2008. "Who drives divergence? Identity signaling, outgroup dissimilarity, and the abandonment of cultural tastes." *Journal of Personality and Social Psychology* 95:593–607.

58. "There Are So Many Kinds of Yoga. This Chart Can Help." 2013. *Huffpost Healthy Living*, September 16. www.huffingtonpost.com/2013/09/16/yoga-chart -infographic_n_3915189.html.

7. REBEL COOL

1. Gay, Peter. 2008. *Modernism: The Lure of Heresy*. New York: W. W. Norton.

2. Mailer, Norman. 1957. "The White Negro: Superficial Reflections on the Hipster." *Dissent*, Fall, pp. 276–93.

3. Mailer, "White Negro."

4. Freud, Sigmund. (1908). "'Civilized' sexual morality and modern nervous illness." In J. Strachey, ed. *The Standard Edition of the Complete Psychological Works of Sigmund Freud*, vol. 9. London: Hogarth Press, pp. 181–204.

5. In the context of Freud's theory of sexuality, see Murphy, T. F. 1983. "Freud reconsidered: Bisexuality, homosexuality, and moral judgement." *Journal of Homosexuality* 9:65–77.

6. Mailer, 1957, "White Negro."

7. Lindner, Robert. 1956. *Must You Conform?* New York: Rinehart.

8. Halmos, Paul. 1957. *Towards a Measure of Man: The Frontiers of Normal Adjustment*. New York: Routledge & Kegan Paul, p. 90.

9. Lindner, Robert. 1954. "Rebels or Psychopaths?" *Time*, December 6.

10. Lindner, 1956, *Must You Conform?*

11. Black, Donald W. 2013. *Bad Boys, Bad Men: Confronting Antisocial Personality Disorder (Sociopathy)*. New York: Oxford University Press.

12. Klabbers, G., et al., 2009. "Measuring rebelliousness and predicting health behaviour and outcomes: An investigation of the construct validity of the social reactivity scale." *Journal of Health Psychology* 14:771–79.

13. Majors, Richard, and Janet Mancini Billson. 1992. *Cool Pose: The Dilemmas of Black Manhood in America*. New York: Jossey-Bass.

14. Stearns, Peter N. 1994. *American Cool: Constructing a Twentieth-Century Emotional Style*. New York: NYU Press.

15. Webster, Richard. 1995. *Why Freud Was Wrong: Sin, Science, and Psychoanalysis*. New York: Basic Books.

16. Wylie, Philip. 1942. *Generation of Vipers*. New York: Pocket Books (1955 ed.), p. 185.

17. Graham quoted in Whitfield, Stephen J. *The Culture of the Cold War* (2nd ed.). Baltimore, MD: Johns Hopkins University Press, 1996.

18. Schlesinger, Arthur. 2008. *The Politics of Hope* and *The Bitter Heritage: American Liberalism in the 1960s*. Princeton, NJ: Princeton University Press, p. 301.

19. Riesman, David. 1961. *The Lonely Crowd*. New Haven, CT: Yale University Press.

20. Fraiman, Susan. 2002. *Cool Men and the Second Sex*. New York: Columbia University Press, p. xv.

21. *The Complete Letters of Sigmund Freud to Wilhelm Fliess, 1887–1904*. 1985. Jeffrey Moussaieff Masson, trans. Cambridge, MA: Belknap Press.

22. Garber, Marjorie. 1995. *Bisexuality and the Eroticism of Everyday Life*. New York: Routledge, p. 186.

23. Jones, Ernest. 1953. *The Life and Work of Sigmund Freud*, vol. 1. New York: Basic Books, p. 317.

24. Ginsberg, Allen. 1995. *Journals Mid-Fifties 1954–1958*. Gordon Ball, ed. New York: HarperCollins.

25. Burroughs, William. *The Adding Machine*. New York: Grove Press, p. 153.

26. Amburn, E. 1998. *Subterranean Kerouac: The Hidden Life of Jack Kerouac*. New York: St. Martin's Press.

27. Jonason, Peter K., N. P. Li, G. D. Webster, and D. P. Schmitt. 2009. "The dark triad: Facilitating a short-term mating strategy in men." *European Journal of Personality* 23:5–18.

28. Hefner quoted in Watts, Steven. 2008. *Mr. Playboy: Hugh Hefner and the American Dream*. Hoboken, NJ: Wiley, p. 134.

29. Amburn, 1998, *Subterranean Kerouac*; Bast, William. 2006. *Surviving James Dean*. New York: Barricade Books.

30. Ehrenreich, Barbara. 1983. *The Hearts of Men: American Dreams and the Flight from Commitment*. New York: Anchor Books.

31. D'Emilio, John, and Estelle B. Freedman. 1997. *Intimate Matters: A History of Sexuality in America*. Chicago: University of Chicago Press.

32. Kaplan, Fred. 2009. *1959: The Year Everything Changed*. Hoboken, NJ: Wiley.

33. Gair, 2007, *American Counterculture*, p. 4.

34. For a review and assessment, see Scott, Isabel M. L., Andrew P. Clark, Lynda G. Boothroyd, and Ian S. Penton-Voak. 2013. "Do men's faces really signal heritable immunocompetence?" *Behavioral Ecology* 24:579–89. For a related theory of male facial signaling, see Fink, B., N. Neave, and H. Seydel. 2007. "Male facial appearance signals physical strength to women." *American Journal of Human Biology* 19:82–87.

35. DeBruine, L. M., B. C. Jones, J. R. Crawford, L.L.M. Welling, and A. C. Little. 2010. "The health of a nation predicts their mate preferences: Cross-cultural variation in women's preferences for masculinized male faces." *Proceedings of the Royal Society B: Biological Sciences* 277:2405–10.

36. Durante, K. M., and N. P. Li. 2009. "Oestradiol level and opportunistic mating in women." *Biology Letters* 5:179–82.

37. Booth, Alan, and James M. Dabbs, Jr. 1993. "Testosterone and men's marriages." *Social Forces* 72:463–77.

38. Fleming, A. S., C. Corter, J. Stallings, and M. Steiner. 2002. "Testosterone and prolactin are associated with emotional responses to infant cries in new fathers." *Hormones and Behavior* 42:399–413.

39. Durante, Kristina M., Vladas Griskevicius, Sarah E. Hill, Carin Perilloux, and Norman P. Li. 2010. "Ovulation, female competition, and product choice: Hormonal influences on consumer behavior." *Journal of Consumer Research* 37: 921–34.

40. Haselton, Martie G., and Steven W. Gangestad. 2006. "Conditional expression of women's desires and men's mate guarding across the ovulatory cycle." *Hormones and Behavior* 49:509–18.

41. Durante, K. M., N. P. Li, and M. G. Haselton. 2008. "Changes in women's choice of dress across the ovulatory cycle: Naturalistic and laboratory task-based evidence." *Personality and Social Psychology Bulletin* 34:1451–60.
42. Beall, A. T., and J. L. Tracy. 2013. "Women are more likely to wear red or pink at peak fertility." *Psychological Science* 24:1837–41.
43. Haselton and Gangestad, 2006, "Conditional expression of women's desires and men's mate guarding"; Haselton, Martie G., Mina Mortezaie, Elizabeth G. Pillsworth, April Bleske-Rechek, and David A. Frederick. 2007. "Ovulatory shifts in human female ornamentation: Near ovulation, women dress to impress." *Hormones and Behavior* 51:40–45.

8. DOTCOOL

1. The group was located at the University of Birmingham.
2. Nuzum, Eric. 2001. *Parental Advisory: Music Censorship in America.* New York: HarperCollins.
3. Record Labeling Hearing Before the Committee on Commerce, Science and Transportation, United States Senate. Susan Baker testimony accessed. www .joesapt.net/superlink/shrg99-529/p12.html.
4. *Nightline* with Ted Koppel, September 13, 1985, www.youtube.com/watch ?v=FarkwYDir2Y.
5. *Frontline* interview with John Warnecke, www.pbs.org/wgbh/pages/frontline /shows/choice2000/gore/warnecke.html.
6. The 1986 obscenity trial of Jello Biafra, lead singer of the punk band Dead Kennedys, in response to complaints from the PMRC heightened the specter of censorship and authoritarian control of musical expression, although Biafra would be acquitted. A few years later, the PMRC turned their attention to rap groups, such as 2 Live Crew.
7. Azerrad, Michael. 1993. *Come As You Are: The Story of Nirvana.* New York: Main Street Books; True, Everett. 2007. *Nirvana: The Biography.* New York: Da Capo Press.
8. Azerrad, 1993, *Come As You Are*, p. 213.
9. Reynolds, Simon. 1991. "Recording View; Boredom + Claustrophobia + Sex = Punk Nirvana." *The New York Times*, November 24, www.nytimes.com /1991/11/24/arts/recording-view-boredom-claustrophobia-sex-punk-nirvana.html.
10. Azzerad, 1993, *Come As You Are*; True, 2007, *Nirvana.*
11. Frank, Thomas. 1997. *The Conquest of Cool.* University of Chicago Press. As Frank notes, Wolfe believed Kesey's counterculture esthetic derived from consumerism. So the theory of co-option (or co-optation) may not get the direction of causation right.
12. "1994: The 40 Best Records from Mainstream Alternative's Greatest Year." 2014. *Rolling Stone*, April 17. www.rollingstone.com/music/lists/1994-the-40 -best-records-from-mainstream-alternatives-greatest-year-20140417.
13. Squires, Nick. 2010. "Vatican Picks the Beatles and Oasis Among Its Top Ten Albums." *The Telegraph*, February 14. www.telegraph.co.uk/news/worldnews /europe/vaticancityandholysee/7236258/Vatican-picks-the-Beatles-and-Oasis -among-its-top-ten-albums.html.

14. There are apparent internal inconsistencies in Frank's argument. At times, he claims that postwar American capitalism was as dynamic a force as revolutionary youth movements. However, he claims it had no transformative power, hence the co-optation theory.

15. For a review, see Sassatelli, 2007, *Consumer Culture.*

16. Whiting, M., and F. Sagne. 2005. "Windows on the World. How the Study of Personal Web Pages Can Provide Insights to Build Brand Strategy." In ESOMAR Innovate Conference Papers, 2005.

17. "Cognac Is the Drink That's Drank by Gs: The Clash of Tradition and US Hip-Hop." 2010. *Cognac Expert* (blog), July 30. http://blog.cognac-expert.com /clash-cultures-traditional-cognac-us-hiphop-america/.

18. Muggleton, David. 2000. *Inside Subculture: The Postmodern Meaning of Style.* Oxford, UK: Bloomsbury Academic.

19. Hebdige, Dick. 1979. *Subculture: The Meaning of Style.* London: Methuen.

20. Bennett, Andy. 1999. "Subcultures or neo-tribes? Re-thinking the relationship between youth, style and musical taste." *Sociology* 33:599–617.

21. As we have been stressing, a status hierarchy is a consensually agreed-upon linear ordering along a unidimensional scale, resulting in such structures as a class stratification. Intense social differentiation may occur in a postindustrial society without such ordering (complex classless inequality). See, for example, Pakulski, Jan, and Malcolm Waters. 1996. *The Death of Class.* Thousand Oaks, CA: Sage Publishers; Pakulski, Jan. 2005. "Foundations of a post-class analysis." In Erik Olin Wright, ed., *Approaches to Class Analysis.* Cambridge, UK: Cambridge University Press, pp. 152–79; Weeden, Kim A., and David B. Grusky. 2012. "The Three Worlds of Inequality." *American Journal of Sociology* 117:1723–85. The relationship between hierarchical vs. horizontal group relations, on the one hand, and political polarization, on the other, is a complex one, but increased egalitarianism may increase, rather than decrease, social conflict. See Nozick, 1974, *Anarchy, State, and Utopia;* Ho, A. K., et al., 2012, "Social dominance orientation"; Guimond, Serge, Richard J. Crisp, Pierre De Oliveira, Rodolphe Kamiejski, Nour Kteily, Beate Kuepper, Richard N. Lalonde, et al. 2013. "Diversity policy, social dominance, and intergroup relations: Predicting prejudice in changing social and political contexts." *Journal of Personality and Social Psychology* 104:941–58.

22. Pew Research Center. 2012. "'Nones' on the Rise: One-in-Five Adults Have No Religious Affiliation." www.pewforum.org/files/2012/10/NonesOnTheRise-full .pdf; Goodstein, Laurie. 2012. "Percentage of Protestant Americans Is in Steep Decline, Study Finds." *The New York Times,* October 9. www.nytimes.com /2012/10/10/us/study-finds-that-percentage-of-protestant-americans-is -declining.html.

23. Pew Research Center. 2014. "Political Polarization in the American Public." www.people-press.org/2014/06/12/political-polarization-in-the-american -public/.

24. Bishop, Bill. 2009. *The Big Sort: Why the Clustering of Like-Minded America Is Tearing Us Apart.* New York: Houghton Mifflin Harcourt.

25. Prometheus is a figure frequently visited, particularly by the Romantics, from Beethoven to Mahler in music and from Goethe to Nietzsche in literature.

26. Cultural critics such as Lasn frequently invoke the heroic rebel image in their work. Lasn, for example, has a slick online booklet cajoling economics students to rebel against the old-guard neoclassical economics professors he likens to a police state and take the "riskier, more exciting path." Be an "agitator, a provocateur, a meme warrior" (by posting one of Lasn's posters on campus).

27. Clark, Dylan. 2003. "The Death and Life of Punk, The Last Subculture." In David Muggleton and Rupert Weinzierl, eds., *The Post-Subcultures Reader.* Oxford, UK: Berg, pp. 223–36.

28. Roderick, John. 2013. "Punk Rock Is Bullshit: How a Toxic Social Movement Poisoned Our Culture." *Seattle Weekly News,* March 6. www.seattleweekly.com/2013-03-06/music/punk-rock-is-bullshit/.

29. Graeber, David. 2011. "Occupy Wall Street's Anarchist Roots." *Aljazeera,* November 20. www.aljazeera.com/indepth/opinion/2011/11/2011112872835904508.html.

30. Graeber, David. 2014. "Savage capitalism is back—and it will not tame itself." *The Guardian,* May 30. www.theguardian.com/commentisfree/2014/may/30/savage-capitalism-back-radical-challenge.

31. Liu, Catherine. 2011. *American Idyll: Academic Antielitism as Cultural Critique.* Iowa City: University of Iowa Press.

32. Gutfeld, Greg. 2014. *Not Cool: The Hipster Elite and Their War on You.* New York: Crown Forum.

33. We should note that in addition to its contribution to the computer revolution, the counterculture obviously changed society in other ways, particularly regarding civil rights. These reforms, however, fall short of revolution, and so reflect co-option.

34. Brand, Stewart. 1995. "We Owe It All to the Hippies." *Time,* March 1. http://members.aye.net/~hippie/hippie/special.htm.

35. Isaacson, Walter. 2011. *Steve Jobs.* New York: Simon & Schuster.

36. For an extensive history of counterculture and the computer revolution, see Markoff, John. 2005. *What the Dormouse Said: How the Sixties Counterculture Shaped the Personal Computer Industry.* New York: Viking; and Turner, Fred. 2006. *From Counterculture to Cyberculture: Stewart Brand, the Whole Earth Network, and the Rise of Digital Utopianism.* Chicago: University of Chicago Press.

37. The intellectual property rights of FOSS are complex. Our definition follows the GPL license.

38. Linux Frequently Asked Questions. www.getgnulinux.org/en/linux/linux_faq/.

39. See National Democratic Institute. "Democracy and Technology." https://www.ndi.org/democracy-and-technology.

40. Goldin, Claudia, and Lawrence F. Katz. 2008. *The Race Between Education and Technology.* Cambridge, MA: Belknap Press.

41. Autor, D. H. 2014. "Skills, education, and the rise of earnings inequality among the 'other 99 percent.'" *Science* 344:843–51.

42. Acemoglu, Daron, Ufuk Akcigit, and Murat Alp Celik. 2014. "Young, Restless and Creative: Openness to Disruption and Creative Innovations." NBER Working Paper no. 19894.

43. Halmos, 1957, *Measure of Man,* p. 90.

44. Schatz, Robin. 2014. "How to Attract the Unconventional Worker." *Inc.*, www
.inc.com/magazine/201406/robin-schatz/how-to-attract-unconventional-
employees.html.

45. The innovation literature has also created much controversy regarding disruptive innovations. See Lepore, Jill. 2014. "The Disruption Machine." *The New
Yorker,* June 23.

46. Sgourev, S. V. 2014. "How Paris gave rise to Cubism (and Picasso): Ambiguity
and fragmentation in radical innovation." *Organization Science* 24:1601–17.

47. Shontell, Alyson. 2012. "The 15 Coolest Offices." *Business Insider,* January 13.
www.businessinsider.com/15-coolest-offices-in-tech-2012-1?op=1.

48. See, for example, the British Council, *Creative Cities*, http://creativecities.brit
ishcouncil.org/creative-industries, and the EU's Creative Cities Project, www
.creativecitiesproject.eu/en/index.shtml.

49. Moretti, 2012. *The New Geography of Jobs*. New York: Houghton Mifflin
Harcourt.

50. For a video representation of jobs lost and gained between 2004 and 2013, see
http://tipstrategies.com/geography-of-jobs/.

51. Moretti, 2012, *New Geography of Jobs*.

52. Gill, Rosalind. 2002. "Cool, creative and egalitarian? Exploring gender in
project-based new media work in Europe." *Information, Communication & Society* 5:70–89.

53. http://protomag.com/assets/the-robot-surgeon; http://blog.notdot.net/tag/damn
-cool-algorithms; www.makeuseof.com/tag/7-cool-html-effects-that-anyone-can
-add-to-their-website-nb/; www.zdnet.com/10-cool-android-apps-to-start-the-year
-7000025635/; www.pinterest.com/mymodernmet/cool-diy-projects/; http://
mashable.com/2014/03/05/boston-startups/.

54. Westcott, Kathryn. 2012. "Are 'Geek' and 'Nerd' Now Positive Terms?" *BBC
News Magazine.* November 15. www.bbc.com/news/magazine-20325517.

55. Chaney, Jen. 2013. "ThinkGeek: The Nerd Company at a Crossroads." *The
Washington Post.* December 12. www.washingtonpost.com/lifestyle/magazine
/thinkgeek-the-nerd-company-at-a-crossroads/2013/12/11/c7d579ba-4b12
-11e3-9890-a1e0997fb0c0_story.html.

56. Green, Emma. 2103. "Mark Zuckerberg: Neither Republican, Democrat, nor
Cool." *The Atlantic,* September 18. www.theatlantic.com/technology/archive
/2013/09/mark-zuckerberg-neither-republican-democrat-nor-cool/279809/.

57. As in such books as Romero, Eric. 2012. *Compete Outside the Box: The Unconventional Way to Beat the Competition.* N.p.: KMFA Press.

58. Sengupta, Somini. 2012. "Why Is Everyone Focused on Zuckerberg's Hoodie?"
The New York Times, May 11. http://bits.blogs.nytimes.com/2012/05/11/why-is
-everyone-focused-on-zuckerbergs-hoodie/?_php=true&_type=blogs&_r=0.

59. See the following websites: http://gizmodo.com/5653143/geeks-versus-hipsters;
www.becomecareer.com/geeks-hipsters; https://blog.blogthings.com/2013/07/24
/new-quiz-are-you-a-geek-or-a-hipster/.

60. Dar-Nimrod, Ilan, et al. 2012. "Coolness: An empirical investigation." *Journal of
Individual Differences* 33:175–85.

61. www.reuters.com/article/2011/09/06/us-americans-cool-survey-idUSTRE
7852H320110906.

62. www.youtube.com/watch?v=PulUKsICY9o.

63. According to Google Trends.

64. Public Policy Polling, May 13, 2013. www.publicpolicypolling.com/PPP_Release _Hipsters_051313.pdf.

65. "Fire Brigade Called Out to Remove Dubai Hipster from Tight Red Pants." 2012. *The Pan-Arabia Inquirer,* July 21. www.panarabiaenquirer.com/wordpress /fire-brigade-called-out-to-remove-hipster-from-tight-red-trousers/.

66. Meister, Jeanne. 2012. "Job Hopping Is the 'New Normal' for Millennials." *Forbes,* August 14. www.forbes.com/sites/jeannemeister/2012/08/14/job-hopping-is-the -new-normal-for-millennials-three-ways-to-prevent-a-human-resource -nightmare/.

67. Castells, Manuel. 1996. *The Rise of the Network Society, The Information Age: Economy, Society and Culture Vol. I.* Cambridge, MA; Oxford, UK: Blackwell; Castells, Manuel. 1997. *The Power of Identity, The Information Age: Economy, Society and Culture Vol. II.* Cambridge, MA; Oxford, UK: Blackwell; Castells, Manuel. 1998. *End of Millennium, The Information Age: Economy, Society and Culture Vol. III.* Cambridge, MA; Oxford, UK: Blackwell.

68. Davenport, Thomas. 2005. *Thinking for a Living.* Boston: Harvard Business Review Press.

69. Arbesman, Samuel. 2012. *The Half Life of Facts.* New York: Current Hard-cover.

70. See the College Board at https://www.collegeboard.org/delivering-opportunity /sat/redesign.

71. For example, Heath and Potter consider contemporary cool as largely conspicuous signaling according to a hierarchy now defined by cool. Heath, Joseph, and Andrew Potter. 2004. *Nation of Rebels: Why Counterculture Became Consumer Culture.* New York: HarperBusiness.

72. Berger, Jonah, and Morgan Ward. 2010. "Subtle Signals of Inconspicuous Consumption." *Journal of Consumer Research* 37:555–69.

73. Manlow, Veronica. 2009. *Designing Clothes: Culture and Organization of the Fashion Industry.* New York: Transaction Publishers.

74. Ruffle, B. J., and Richard Sosis. 2007. "Does it pay to pray? Costly ritual and cooperation." *The BE Journal of Economic Analysis & Policy* 7:Iss.

75. See Featherstone, Mike. 1991. *Consumer Culture and Postmodernism.* London: Sage; McIntyre, R. 1992. "Consumption in contemporary capitalism: Beyond Marx and Veblen." *Review of Social Economy*: 50:50–57.

76. Moretti, 2012, *New Geography of Jobs.*

77. Duncan, Fiona. 2014. "Normcore: Fashion for Those Who Realize They're One in 7 Billion." *New York*, February 26. http://nymag.com/thecut/2014/02/normcore -fashion-trend.html.

78. Duncan, 2014, "Normcore."

79. Merelli, Annalisa. 2014. "A Brief History of Normcore and Other Things That Weren't Things Before They Became Things." *Quartz*, April 22. http://qz.com /201413/a-brief-history-of-normcore-and-other-things-that-werent-things -before-they-became-things/.

80. Frank, Thomas. 2014. "Hipsters, They're Just Like Us! 'Normcore,' Sarah Palin, and the GOP's Big Red State Lie." *Salon*, April 27, www.salon.com/2014/04/27

/hipsters_they%E2%80%99re_just_like_us_normcore_sarah_palin_and_the
_gops_big_red_state_lic/.

81. "President Normcore: Obama Shops at Midtown Gap." 2014. *Gothamist*, March 3. http://gothamist.com/2014/03/11/obama_does_normcore_president _shops.php#.

82. Purser, Craig. 2014. "Beer Industry Numbers Tell the Story of Effective Laws." *Roll Call*, April 29. www.rollcall.com/news/beer_industry_numbers_tell_the _story_of_effective_laws_commentary-232381-1.html.

83. Schwartz, Barry. 2003. *The Paradox of Choice*. New York: Ecco.

84. Purser, 2014, "Beer Industry Numbers."

85. For reviews and some controversies surrounding imitation learning among humans and primates, see Tomasello, Michael, Ann Cale Kruger, and Hilary Horn Ratner. 1993. "Cultural learning." *Behavioral and Brain Sciences* 16:495–511; Byrne, Richard W., and Anne E. Russon. 1998. "Learning by imitation: A hierarchical approach." *Behavioral and Brain Sciences* 21:667–84; Horowitz, Alexandra C. 2003. "Do humans ape? Or do apes human? Imitation and intention in humans (*Homo sapiens*) and other animals." *Journal of Comparative Psychology* 117:325–36.

86. Hayden, Casey, and Mary King. 1965. "Sex and Caste: A Kind of Memo from Casey Hayden and Mary King to a Number of Other Women in the Peace and Freedom Movements." Reprinted at CWLU Herstory Website Archive. http:// uic.edu/orgs/cwluherstory/CWLUArchive/memo.html.

87. See, for example, Gerhard, Jane. 2001. *Desiring Revolution: Second-Wave Feminism and the Rewriting of American Sexual Thought, 1920 to 1982*. New York: Columbia University Press.

88. See, for example, Paglia, Camille. 1990. "Madonna—Finally, a Real Feminist," *The New York Times*, December 14. www.nytimes.com/1990/12/14/opinion /madonna-finally-a-real-feminist.html.

89. Dragani, Rachelle. 2010. "Most Powerful Women of the Past Century." *Time*, November 18. http://content.time.com/time/specials/packages/article /0,28804.2029774_2029776,00.html.

90. Greenburg, Zack. 2013. "The World's Highest-Paid Musicians 2013." *Forbes*, November 19, www.forbes.com/sites/zackomalleygreenburg/2013/11/19/the -worlds-highest-paid-musicians-2013/.

91. Valenti, Jessica. 2009. *The Purity Myth: How America's Obsession with Virginity Is Hurting Young Women*. New York: Seal Press.

92. See "The Silver Ring Thing" website: www.silverringthing.com/tourtheme.asp.

93. Kerner, Ian. 2009. "Can You (and Should You) Have Sex Like a Man?" *The Today Show Health*, February 19. www.today.com/id/29282186/ns/today-today health/t/can-you-should-you-have-sex-man/#.U1lyM_ldXbw.

94. White, Ruth. 2011. "No Strings Attached Sex (NSA): Can Women Really Do It?" *Psychology Today*, November 20. www.psychologytoday.com/blog/culture -in-mind/201111/no-strings-attached-sex-nsa-can-women-really-do-it.

95. Borrow, Amanda P., and Nicole M. Cameron. 2012. "The role of oxytocin in mating and pregnancy." *Hormones and Behavior* 61:266–76.

96. Scheele, Dirk, et al. 2013. "Oxytocin enhances brain reward system responses in men viewing the face of their female partner." *Proceedings of the National Academy of Sciences* 110:20308–13.

97. Vrangalova, Zhana. 2014. "Does casual sex harm college students' well-being? A longitudinal investigation of the role of motivation." *Archives of Sexual Behavior* doi:10.1007/s10508-013-0255-1.

98. Kimmel, Michael. 2014. "How 'Free to Be' Heralded the Most Successful Revolution of Our—or Any—Era." *Huffington Post*, June 8. www.huffington post.com/michael-kimmel/how-free-to-be-heralded-the-most-successful -revolution_b_5097715.html. See also Kimmel, Michael. 2012. *Manhood in America: A Cultural History* (3rd ed.). New York: Oxford University Press.

99. U.S. Department of Labor. 2011. "Women's Employment During the Recovery." www.dol.gov/_sec/media/reports/femalelaborforce/.

100. In some sectors, however, such as software engineering, there remains a low rate of females.

101. Leblanc, Lauraine. 1999. *Pretty in Punk: Girls' Gender Resistance in a Boys' Subculture*. New Brunswick, NJ: Rutgers University Press.

102. Public Policy Polling, May 13, 2013. www.publicpolicypolling.com/PPP _Release_Hipsters_051313.pdf.

103. *The Hipster Mom*. www.thehipstermom.com/.

104. Barber, Benjamin. 2007. *Consumed: How Markets Corrupt Children, Infantilize Adults, and Swallow Citizens Whole*. New York: W. W. Norton.

105. Barber, Benjamin. 2007. "Spent Youth." *The American Conservative*, May 7. www.theamericanconservative.com/articles/spent-youth/.

106. See www.ted.com/talks/daphne_bavelier_your_brain_on_video_games. For research, see Bavelier, D., C. S. Green, P. Schrater, and A. Pouget. 2012. "Brain plasticity through the life span: Learning to learn and action video games." *Annual Reviews of Neuroscience*, 35:391–416; Bavelier, D., C. S. Green, D. H. Han, P. F. Renshaw, M. M. Merzenich, and D. A. Gentile. 2011. "Brains on video games." *Nature Reviews Neuroscience*, 12:763–68; Dye, M.G.W., C. S. Green, and D. Bavelier. 2009. "The development of attention skills in action video game players." *Neuropsychologia* 47:1780–89.

107. Schwartz, Barry, 2005. "The Paradox of Choice." www.ted.com/talks/barry _schwartz_on_the_paradox_of_choice.

108. Iyengar, Sheena S., and Mark R. Lepper. 2000. "When choice is demotivating: Can one desire too much of a good thing?" *Journal of Personality and Social Psychology* 79:995–1006.

109. Soble, Jeffrey A. 2013. "Car Dealers Lose Negotiating Leverage & Profits . . . and Like It." *Dashboard Insights*, November 11. www.autoindustrylaw blog.com/2013/11/11/car-dealers-lose-negotiating-leverage-profitsand-like-it/.

110. Quartz and Sejnowski, 1997, "Neural basis of cognitive development."

111. Wampole, Christy. 2012. "How to Live Without Irony." *The New York Times*. November 17. http://opinionator.blogs.nytimes.com/2012/11/17/how -to-live-without-irony/.

112. https://www.youtube.com/watch?v=Mc335NvEJ_0.

113. See, for example, Howe, Neil, and William Strauss. 2000. *Millennials Rising: The Next Generation*. New York: Vintage.

114. https://www.youtube.com/watch?v=32LCwZFoKio.

115. Chaplin, Julia. 2003. "A Hat That's Way Cool. Unless, of Course, It's Not." *The New York Times*, May 18. www.nytimes.com/2003/05/18/fashion/18HATS .html.

116. Fitzgerald, Jonathan. 2012. "Sincerity, Not Irony, Is Our Age's Ethos." *The At-lantic*, November 20. www.theatlantic.com/entertainment/archive/2012/11/sinc erity-not-irony-is-our-ages-ethos/265466/.

117. Breen, T. H. 2004. *The Marketplace of Revolution: How Consumer Politics Shaped American Independence.* Oxford, UK: Oxford University Press.

118. Stolle, Dietlind, and Michele Micheletti. 2013. *Political Consumerism: Global Responsibility in Action.* Cambridge, UK: Cambridge University Press.

119. Willis, M. M., and J. B. Schor. "Does Changing a Light Bulb Lead to Changing the World? Political Action and the Conscious Consumer." *The Annals of the American Academy of Political and Social Science* 644:160–90.

120. Monbiot, George. 2010. "The Values of Everything." www.monbiot.com/2010 /10/11/the-values-of-everything/.

121. Sexton, Steven E., and Alison L. Sexton. 2014. "Conspicuous conservation: The Prius halo and willingness to pay for environmental bona fides." *Journal of Environmental Economics and Management* 67:303–17.

122. Crompton, Tom. 2010. "Common Cause: The Case for Working with Our Cultural Values." http://assets.wwf.org.uk/downloads/common_cause_report .pdf.

123. Bénabou, Roland, and Jean Tirole. 2006. "Incentives and prosocial behavior." *American Economic Review* 96:1652–78.

124. Bénabou and Tirole, 2006, "Incentives and prosocial behavior."

125. Griskevicius, Vladas, Joshua M. Tybur, and Bram Van den Bergh. "Going green to be seen: Status, reputation, and conspicuous conservation." *Journal of Personality and Social Psychology* 98:392–404.

126. Sexton and Sexton, 2014, "Conspicuous conservation."

127. Sexton and Sexton, 2014, "Conspicuous conservation." We should mention that this discussion of the Prius makes no claims about whether the Prius is in fact the most optimal choice in terms of environmental impact, or about the extent to which Prius ownership does or does not impact climate change. What is revealing about the pattern of Prius ownership is the fact that consumption can be a prosocial and altruistic signal. Indeed, a lesson from this in the form of a policy recommendation is to convert private consumption choices into public ones, such as making household energy consumption public knowledge. When this is done, conspicuous conservation will motivate consumers. Monetary subsidies should then be used primarily where energy consumption cannot be made public, since private consumption is not affected by signaling and esteem motives.

128. Dietz, Thomas, et al. 2009. "Household Actions Can Provide a Behavioral Wedge to Rapidly Reduce US Carbon Emissions." *Proceedings of the National Academy of Sciences of the United States of America* 106:18452–56; Girod, Bastien, Detlef Peter van Vuuren, and Edgar G. Hertwich. 2014. "Climate policy through changing consumption choices: Options and obstacles for reducing greenhouse gas emissions." *Global Environmental Change* 25:5–15.

129. See, for example, the Ellen MacArthur Foundation, www.ellenmacarthurfoun dation.org/.

130. McDonough, William, and Michael Braungart. 2002. *Cradle to Cradle: Remaking the Way We Make Things.* New York: Macmillan.

131. For more information on the Cradle to Cradle Products Innovation Institute, cradle-to-cradle certification, and a list of certified products, see 222.c2certified.org.

132. See, for example, Princen, Thomas, Michael Maniates, and Ken Conca, eds. 2002. *Confronting Consumption*. Cambridge, MA: MIT Press.

133. See, for example, Diamandis, Peter, and Steven Kotler. 2012. *Abundance: The Future Is Better Than You Think*. New York: Free Press.

ACKNOWLEDGMENTS

This book owes its existence to an inquiry we received from Lisa Ling in early 2004. Her questions got us thinking about how cool impacts the brain and led us to think seriously about cool's economic role. We never would have guessed that her query would be the impetus for a decade of research, but the remarkable thing about science is the unexpected journeys it sometimes takes us on. So our thanks to Lisa for sparking our curiosity and prompting us to ask the questions that have led us here, to this book.

We have also had the good fortune to investigate these questions while surrounded by an incredible group of researchers. We thank the members of the Quartz lab—Cedric Anen, Ulrik Beierholm, Tony Bruguier, Shreesh Mysore, and Kerstin Preuschoff, along with Ming Hsu—for creating an exciting intellectual environment. At Caltech, we have also been fortunate to be among an extraordinary group of researchers, including Peter Bossaerts, Colin Camerer, Antonio Rangel, John Allman, Ralph Adolphs, John O'Doherty, Fiona Cowie, Chris Hitchcock, and Gideon Manning. We would also like to thank Ralph Miles and Stephen Flaherty for their assistance with the brain-imaging experiments, and Read Montague and members of his lab for many interactions over the years relating to decision making and the brain.

We would like to thank the design students at Art Center College of Design who helped us with our experiments. We have also benefited from our many interactions with designers, architects, and others at Art Center. We are grateful to Tim McPartlin at Lieberman Research Worldwide for many discussions and interactions over the years relating to marketing research and the brain. At Farrar, Straus and Giroux, we would like to thank Eric Chinski for helping us shape the first contours of this book many years ago and for his encouragement and patience as we worked to turn those early ideas into a draft. We would also like to thank Amanda Moon for her incredible energy and editorial wizardry and her immense help in turning that draft into a book. We owe thanks to Laird Gallagher for his assistance.

A special thank-you to Katinka Matson for all her help guiding this project over the years.

Steve would like to thank his family and friends for their encouragement. In particular, he'd like to thank his children, Evelyn, Alden, and Elliot, for all their help with this project. Thanks for being the go-to resource for all things Millennial and for your patience when asked at the dinner table such questions as whether LeBron James is a hipster. Steve would like to thank Karen for all the discussions that helped shape the ideas in the book, for her critical feedback, and for being a constant support through it all.

Anette would like to thank her family and friends for all the love, encouragement, and support they gave her. Special thanks go to her sister, Kamilla, who is always there, is an extremely good listener, and is a rock when it comes to good advice. Special thanks also go to Anette's loving and supportive parents, who always understood the importance of education and hard work. Her mother, Janina, is a model of strength, independence, and perseverance, not to mention fashion and style. Her father, John, was always a source of inspiration through his kindness, wisdom, and curiosity about the world, philosophy, and languages. Both have always been tremendously supportive when Anette has been pursuing her dreams.

INDEX